The Space Station

The Space Station

A Personal Journey

Hans Mark

Duke University Press   Durham 1987

©1987 Duke University Press
All rights reserved
Printed in the United States of America
on acid-free paper ∞
Library of Congress Cataloging-in-Publication Data
Mark, Hans Michael, 1929–
The space station.
Bibliography: p.
Includes index.
1. Astronautics and state—United States. 2. Space
stations. I. Title.
TL789.8.U5M27   1987   387.8'0973   86-32892
ISBN 0-8223-0727-8

Contents

Preface   vii

I   Beginnings   1

II   Nuclear Lessons   10

III   Livermore   14

IV   The Space Race Starts   21

V   The Start at Ames and the Space Shuttle   30

VI   The Development of the Shuttle, Colonies in Space, and Politics   47

VII   The Air Force and Space   59

VIII   Space Policy, Arms Control, and Organizational Problems   75

IX   The Shuttle Program Has Problems   90

X   The Election of President Reagan   111

XI   The Effort Starts in Earnest   126

XII   Space Policy and Edwards Air Force Base— July 4, 1982   143

XIII   The *Enterprise* in Europe—May–June 1983   155

XIV   The Final Push   162

XV   The President Decides   188

XVI   The Congressional Debate   197

XVII   Another Beginning   210

XVIII   Tragedy and Tomorrow   215

Appendixes   225

Index   255

*To the memory of the seven space travelers
who died aboard the spaceship* Challenger,
January 28, 1986

*Mr. Francis R. (Dick) Scobee, Commander
Commander Michael J. Smith, U.S. Navy, Pilot
Lt. Colonel Ellison S. Onizuka, U.S. Air Force,
Mission Specialist One
Dr. Judith A. Reisnik, Mission Specialist Two
Dr. Ronald E. McNair, Mission Specialist Three
Ms. S. Christa McAuliffe, Payload Specialist One
Mr. Gregory B. Jarvis, Payload Specialist Two*

Preface

My original purpose in writing this book was to tell the story of how the space shuttle came to be and how the American space station program was initiated. For much of that period (from 1969 to 1984) I had a good vantage point from which to observe the events that led to these developments.

Chapters I through XVII were written in the summer and autumn of 1984 shortly after I made the decision to leave my post as deputy administrator of NASA. I have made no changes in these chapters because I wanted to keep the description of the atmosphere in NASA and the space program at the time, and I also wanted to describe the attitudes I had toward the program before the loss of *Challenger* on January 28, 1986. I thought that these were worth preserving in print.

The last chapter of the book was written after the *Challenger* accident. There is no doubt that this event has, at least temporarily, changed the position of NASA and that it has been a major setback for the American space program. I felt that it would be worthwhile to put down my own reflections on the event and to draw some conclusions about the future of this nation's efforts in space operations when the impact of *Challenger*'s loss is taken into account.

A great many people contributed to the writing of this book. The following individuals reviewed the manuscript and made valuable suggestions: Mr. Robert F. Allnutt, Dr. Roger E. Batzel, Mr. James M. Beggs, Mr. Ben Bova, Dr. Harold Brown, Dr. Larry D. Carver, Mr. Arthur C. Clarke, Mr. Philip E. Culbertson, Mr. James C. Elms, Dr. Maxime A. Faget, Dr. Terence T. Finn, Dr. James C. Fletcher, Dr. Sylvia D. Fries, Mr. Robert R. Gilruth, Professor James E. Katz,

Mr. John G. Kester, Ms. Dorothy J. Kokoski, Professor Herman F. Mark, Mrs. Heather Moore, Mr. C. Thomas Newman, Colonel Gilbert D. Rye, U.S. Air Force (retired), Senator Harrison H. Schmitt, Mr. G. Harry Stine, Mr. Norman Terrell, Mrs. Josephine L. Watson, and Mr. John F. Yardley. I would like to express my sincere thanks to all of these people; the book is much better because of the time they spent making many comments and corrections.

I am particularly grateful to Ms. Dorothy J. Kokoski who typed the first draft of the manuscript and to Mrs. Sheila Simmons who did such an excellent job on the final draft. Mrs. Becky Boyer and Mrs. Sylvia Hodges also helped with the typing. The editor is a most important person in any enterprise such as this one, and I am grateful for the intelligence and the diligence that Mr. Reynolds Smith of the Duke University Press brought to this effort. Finally, I am grateful for the encouragement of my wife, Dr. Marion T. Mark, who had the idea that I should keep a daily diary when we moved to Washington in 1977. If she had not made that suggestion, this book could not have been written.

Hans Mark
Austin, Texas
January, 1987

Beginnings

Books are important, and they can influence lives. I want to begin this story by talking about a book I received as a present on my twelfth birthday in 1941. It is a remarkable book about space travel that was written by an Englishman, P. E. Cleator, who was for many years a leader of the British Interplanetary Society. (Cleator served as the first president of the society, having been elected in October 1933.) The book was published in 1936 and is titled *Rockets Through Space*. (It also carries the subtitle *The Dawn of Interplanetary Travel*.) The book contains a good popular description of the state of the art in rocket technology in 1935 (the year the book was written), and it also includes a good technical summary of what could be done in terms of space travel once the large rockets that are envisioned in the work were built. In these sections, space stations (or outward stations as Cleator called them) and future trips to other planets were prominently mentioned.

All of this was pretty heady stuff for a twelve year old. At first, of course, I did not really understand what the technical issues were, but I kept rereading the book over the years. Slowly it dawned on me that Cleator and his colleagues were on to something really important and that I wanted somehow to be a part of it.

Much of the technical portion of *Rockets Through Space* deals with the early development of liquid-fueled rocket motors. As early as 1923, in his book, *Die Rakete zu den Planetenraumen*, Professor Hermann Oberth, who was one of the pioneers in making quantitative calculations of the performance of rockets, pointed out the advantage of liquid-fueled rockets over those that used solids—gunpowder or nitroglycerine in those days—for propulsion. The essential point was that liquid fuels could be metered into the combustion chamber in such a way that they could be controlled, and this would turn out to be critical in controlling the flight path of the rocket. Furthermore, solid fuels had a tendency to explode violently and were therefore deemed by Oberth to be more dangerous. One of the early rocket pioneers, Reinhold Tilling, was killed in just such an accident. Finally, liquid-fueled rockets were, at least theoretically, more efficient because the combustion products tended to have lower molecular weights and therefore a higher specific impulse.

In addition to his scientific activities, Oberth also participated in the organization of the experimental work that was initiated to prove out his ideas. In June 1927 Oberth, his student, Max Valier, and the engineer, Johannes Winkler, met in the then German city of Breslau (now Wroclaw in Poland) and founded the Verein für Raumschiffahrt, which—loosely translated—meant the Society for Space Flight. The Verein für Raumschiffahrt, or VfR as it was commonly called during the years it was active, played a critical role in the early years of rocket technology. Early on, the membership of the VfR decided to concentrate whatever resources they could muster on the development of liquid-fueled rocket motors. This decision turned out to have far-reaching consequences, not only because it was technically correct, but also because it drew a truly authentic genius, Wernher von Braun, into the world of rocketry. (Von Braun joined the work of the VfR in 1932.)

The VfR decided to build small experimental liquid-fueled rocket motors. These were called MIRAKs (for *Mi*nimum *Rak*ete)—I guess this was the first acronym I ran across in what is now a long career in dealing with those awful things—and they were simply small rocket engines whose performance could be measured on appropriately designed test stands. What really intrigued me about the description of the MIRAKs in the book was not only that there were drawings showing how they were intended to work but also photographs of how they looked when they were running. These lent a sense of reality to the entire enterprise that other science and/or science fiction books I read at the time could not match.

In addition to the technical descriptions of the early experiments on liquid-fueled rockets, there were some grand speculations in Cleator's book. I have already mentioned the "Outward Station" (pp. 171–74) describing the work of the Austrian engineer Count Guido Von Pirquet;

in the following chapter, titled "Achievement and After," Cleator describes what must be done to achieve the objective of interplanetary travel. He discusses the problem of sustaining people in space—an early description of what we today call life support systems—and then goes on to speculate that strange living things might be found on other planets, especially Venus and Mars. I was absolutely enthralled by these visions because the pictures were vividly drawn and there seemed to me to be a thread of technical credibility in the book that made the prospects real enough to touch. On rereading the book recently, I am still struck by the power of Cleator's imaginative words in that most important chapter.

I should not create the impression that *Rockets Through Space* is only a flight of fancy. Quite the contrary; it presented—as I have already said—a very sober technical analysis of the situation as it existed in 1935. The formidable nature of the problem was described by Cleator in a chapter titled "The Problem of Problems: Fuel." In this chapter Cleator calculates what is required to put a spaceship into an escape trajectory, go to a planet, and then return. His estimate was that using liquid oxygen–liquid hydrogen fuel—the most efficient available —a 41,000-ton launch vehicle (including fuel) would be required to make a trip to a planet with a twenty-ton (four-passenger) spaceship and to return it to earth (pp. 66–67). Fortunately, this estimate turned out to be quite pessimistic as subsequent events have proved. During the Apollo program, an 8.2-ton spaceship was sent to the moon and brought back with a rocket—the Saturn V—that had a gross takeoff weight at launch of 3,200 tons. Most of the difference is due to the much stronger and lighter materials available today that permitted a fuel-to-dry-weight ratio of 8.8 for the Saturn V as opposed to about 6.0 for Cleator's rocket. A smaller difference is that the lunar journey required a somewhat smaller escape velocity than the one Cleator assumed would be necessary for one of his interplanetary trips.

Cleator wound up his chapter on "The Problem of Problems" pessimistically, commenting at some length on the formidable difficulties that would be encountered in lifting a ship weighing 40,000 tons off the earth. He says "this is not an impossible size by any means, but the cost of transporting a rocket ship weighing but 20 tons, manned by a crew of only four persons, into space and back, assumes the terrifying total of $100,000,000." In a technological sense Cleator was pessimistic (and realistic), but as a financial forecaster he was the wildest optimist! What was most important from my viewpoint were the last few sentences of this all-important chapter (p. 68):

> The trouble lies in the weakness of the fuels at present available. Without doubt, the fuel problem is the most vexing question with which interplanetary travel is faced at the moment. And it must be confessed that there is no satisfactory substitute for the oxygen-hydrogen mixture yet in sight. But as past history clearly shows, no matter how insoluble problems have at first appeared in the past, unremitting labor and patient research have eventually triumphed. It would be strange indeed if this particular problem proved an exception. And, as we shall see, there are several possible avenues of approach. The most simple and direct method, of course, is the discovery of more powerful fuels. And in this connection, we are in the tantalizing position that even present day fuels possess more than enough energy if only we knew how to release it and use it. Just as molecular energy is so freely used today, so atomic energy may bring interplanetary travel within easy reach tomorrow.

There it was. After discussing the problems with chemical fuels for the entire chapter, Cleator, in the last sentence, proposes that atomic energy (i.e., nuclear energy) might be the answer. There is no doubt at all that this is what he actually had in mind. Later on in the book (pp. 125 and 126) in a chapter titled "Ways and Means," there are two paragraphs that turned out to be of the utmost importance to me. It is worthwhile quoting these paragraphs in full:

> Finally, and no matter how remote or fanciful it may seem at present, there remains the possibility of utilizing atomic energy. Those to whom the idea seems wholly fanciful should remember that in the disintegration of radium, and other radioactive substances, there cannot be the slightest doubt that a slow process of atomic disruption is going on. It is true that the process cannot be accelerated or decelerated by any known means. But the fact remains that nearly 2,000,000 calories of heat are evolved during the degradation of one gram of radium—no less than 250,000

times as much heat as is evolved during the combustion of a similar weight of coal.

There can be no doubt, therefore, but that matter contains tremendous stores of potential energy, if only we knew how to release and use it. The radioactive elements offer no hope at present, for their rate of disintegration is comparatively slow, and is not available for doing useful work. But it is not inconceivable that some means of accelerating their rate of decomposition will ultimately be found. After which, a method of disrupting the most stable elements will probably follow.

Looking back on it now, this is truly a most remarkable statement. Three years before Otto Hahn and Fritz Strassmann (1938) performed their epoch-making experiments demonstrating uranium fission, Cleator speculated on the likelihood that this event would actually occur! Furthermore, he also concluded that nuclear energy might eventually be useful as a source of propulsive power, a really remarkable leap of the imagination.

As an aside, it might be instructive to compare Cleator's view of these possibilities with those held by Lord Rutherford at about the same time. It is ironic that Rutherford, who probably did more than anyone else to clarify our understanding of the structure of the atom and its nucleus, failed to realize the potential of nuclear energy. Writing in 1937, only one year before the crucial experiment of Hahn and Strassmann, he said that "The outlook for gaining useful energy from the atoms by artificial processes of transformation does not look very promising." All of which is a classic example of Arthur C. Clarke's "first law of prophecy," which goes like this: "When a distinguished and elderly scientist says that something within his areas of competence can be done, he is probably right, and when he says that something cannot be done, he is almost certainly wrong."

Among other things, I learned from Cleator's book not to be too trusting of expert opinions—the poets and the dreamers are more often right in the long run than those of us who are too close to technical work.

I was sixteen years old (1945) when the full impact of Cleator's prediction about the potential of nuclear power for propulsion struck home. The first nuclear explosion near Alamogordo, New Mexico (July 1945), and its successors at Hiroshima and Nagasaki the following month, were potent signals for the future. By that time I thought I knew enough to evaluate for myself the technical points that Cleator had made, and I came to the conclusion that rockets capable of putting useful payloads on the moon or on other planets were at least half a century in the future. However, I believed Cleator's hint that nuclear energy might somehow be the eventual answer, and so I decided to study nuclear physics. With luck, I thought, I would live long enough to see the first men on the moon by about the year 2,000. And, I believed, that the best way I could help to hasten the day of this event would be to study nuclear physics and learn how to harness nuclear energy for rocket propulsion.

My decision to study nuclear physics was confirmed two years later (1947) when I was given another book that also had significant influence on my thinking. This one was by the prominent German science writer, Willy Ley, and it was called *Rockets and Space Travel* with the subtitle *The Future of Flight Beyond the Stratosphere*. This book was written at about the same technical level as Cleator's earlier volume, but it had very much more to say. Progress in the decade between 1935 and 1945 was truly fantastic. From the primitive MIRAKs we had come to the V-2s, and from vague speculations about the future of nuclear energy, we had the awesome reality of the atomic bomb.

Ley repeated in his book Hermann Oberth's old calculations and came to the same conclusion reached by Cleator, which was that chemically fueled rockets capable of putting people on the moon and returning them to earth would be impractically large. On pages 281–83 Ley describes what we then knew about uranium fission and speculates on how this new source of energy might be employed to drive spaceships. After discussing some of the obstacles, Ley concludes this section with the following sentence: "The important point is that we now know of a fuel which, theoretically at least, can produce ultra-fast exhaust velocities." As far as I was concerned, that was good enough. As things turned out, Ley's speculation was wrong, but I did not know enough at the time to understand the problem. The achievement of ultra-fast exhaust velocities depends on the temperature of the working fluid in the combustion chamber and is independent of the energy source used to produce that temperature. The maximum temperature depends on the melting point of the materials of which the combustion chamber is made. Nuclear energy sources cannot exceed that temperature without melting the chamber, and, therefore, if conventional rocket technology is employed, nuclear energy is probably no better than other sources for achieving "ultra-fast exhaust velocities." Nuclear power becomes advantageous only when exotic propulsion schemes such as ion propulsion or plasma jets are employed along with nuclear energy.

Curiously enough, I was less interested at the time in the last chapter of Ley's book, which is called the "Terminal in Space." In this section of his book Ley provides a comprehensive description of what we today call the space station. Having just completed his discussion of nuclear energy in which he reached his somewhat erroneous conclusion, he starts the chapter as follows (p. 284):

> Without atomic energy not very many of the goals of rocket research are within direct reach. We can, of course, build high-altitude rockets of all kinds and sizes. We can go out into space, with observers aboard, for several earth diameters. We can send unmanned rockets to the moon and we might toy with the idea of a manned trip around the moon.
>
> Landing on the moon is already beyond the borderline of what chemical fuels can do. And to reach the neighboring planets directly will certainly require the use of atomic energy. But even when atomic energy was still thought to be centuries in the future there existed a theoretical method of landing on the moon and of going to Mars and to Venus with chemical fuels. That theoretical method was a refueling terminal in space, a cosmic stepping stone for spaceships which were too weak to reach another planet directly.

Here then is one of the major arguments for the establishment of a space station. Ley starts the chapter by describing Hermann Oberth's early work and then concentrates on recounting the work of the Austrian Count Guido Von Pirquet. Von Pirquet is the real "father of the space station" since he was the one who made the detailed quantitative calculations in 1928 (see the VfR journal *Die Rakete*, vol. 2, pp. 134–40) to show how the space station can be used as a staging base for more ambitious missions to the moon and to the planets. Later on, Ley foresaw the use of a space station as a laboratory in space and predicted correctly the other uses that have been developed for the space station: The space station would be a laboratory in earth orbit to be used for conducting experiments in physics, astronomy, biology, and earth observations. The space station would be a construction base, that is, as Ley put it: "The lack of gravitational strain not only permits absolutely new experiments of all kinds; it also permits the construction of new types of equipment and instruments." He then went on to describe the construction of a large astronomical telescope in earth orbit. In short, Ley, thirty-seven years ago, correctly assessed the justifications for building a space station. The arguments he made are precisely the ones that were used to persuade President Reagan to include the permanently manned space station in his fiscal year 1985 program.

Ley's book was important in one other respect. Unlike Cleator, Ley gave much space to the work done by the great American rocket pioneer, Dr. Robert H. Goddard. In 1919 Goddard published a seminal paper called "A Method of Researching Extreme Altitudes" in which he reported on a series of meticulously designed experiments to prove the theories of rocket propulsion in a quantitative manner. He used solid-fueled experimental rocket motors for this purpose. Goddard soon turned his attention to liquid-fueled rockets, and, working alone in his laboratory at Clark University in Worcester, Massachusetts, where he was a professor of physics, he succeeded in making a liquid-fueled (gasoline-liquid oxygen) rocket work on November 1, 1923, and he flew one for the first time on March 16, 1926. Goddard was thus well ahead of the work conducted by the Germans of the VfR, where the first static firings of the MIRAKs did not take place until 1929, and the first flights of the Repulsor rockets using motors based on the MIRAK design were carried out two years later.

Goddard was very secretive and worked alone. Nevertheless, there is no doubt that he was an authentic genius who made enormous contributions to the art and science of rocketry. He was especially active in developing guidance and stabilization systems for liquid-fueled rockets that he built and flew in the decade of the 1930s. This work was carried out at the Mescalero Ranch near Roswell, New Mexico, and was funded by a grant from the Guggenheim Foundation.

I would like to end this chapter on a personal note. Robert Goddard was not only a great experimental physicist; he was also a philosopher who understood what the process of science was all about. Speaking at his high school graduation in 1904, Goddard made the famous state-

Herman F. Mark
in a characteristic pose
—about 1956.

ment for which he is still remembered. It is worthwhile to present the entire quotation: "The fact that errors in scientific reasoning are so common should not serve as a discouragement. Every fallacy we detect can show us where we are at fault and guide us toward the truth . . . we should derive courage from the fact that all sciences have been, at some time, in the same condition and that it has often proved true that the dream of yesterday is the hope of today and the reality of tomorrow." When my parents gave me P. E. Cleator's book forty-three years ago, my mother put the date of my birthday at the top of the front page in her handwriting, "June 17, 1941." Below this, my father wrote "The phantasy of today is the reality of tomorrow. For Hans, this book from Herman."

I do not know whether my father had ever seen Goddard's quotation when he wrote the dedication. I do know that I have tried to live by it ever since.

Nuclear Lessons

I graduated from Stuyvesant High School in New York City in the summer of 1947 and decided to attend the University of California in Berkeley to study nuclear physics. Although I was partly influenced by what I had read in the two books I described in the previous chapter, I had other reasons as well. I felt that the prospects of space travel were so far in the future that I wanted to work on something that was of more immediate interest. I guess I was simply not willing to take the risk.

Shortly after the end of the war in 1945, the University of California instituted a regulation that only residents of California could attend. Since I was not a resident, I had a problem; but this was fixed by Professor Joel H. Hildebrand who was an old and close personal friend of my father, Professor Herman F. Mark of the Polytechnic Institute of New York. In due course I was admitted to the university, and I arrived in Berkeley in September 1947, shortly after my eighteenth birthday. In those days Berkeley was the major center of activity in nuclear physics, and Professor Ernest O. Lawrence, who had received the Nobel Prize in physics for his invention of the cyclotron in 1939, was the leading figure of the university's scientific establishment. Professor Hildebrand was kind enough to introduce me to Lawrence soon after I arrived in Berkeley, and it was an awesome moment for me when I was able to shake the hand of this truly great man. Living in Berkeley in those days was a heady experience. Among my teachers in the physics department were several who would, in due course, win their own Nobel prizes, but the one I remember best was Professor August Carl Helmholz, who was my undergraduate advisor. I was very lucky to meet him, and we subsequently became good friends. Professor Harold Lewis was also an influential teacher, and it was he who eventually advised me to go to MIT for my graduate studies. And so, early in 1951 (since I had graduated six months early), together with my new wife, the former Marion G. Thorpe, I set out for Boston.

Next to Berkeley, MIT was probably the most active center in nuclear physics at the time. In contrast to Berkeley, where the work was concentrated around Lawrence's cyclotrons, the nuclear work at MIT was dominated by the electrostatic particle accelerators developed there by Professor Robert J. Van de Graaff. Shortly after arriving at MIT, I went to work in a group headed by Professor Clark D. Goodman. The group was known as the Neutron Physics Group, and the work was funded by the Navy's nuclear submarine propulsion program that was just getting started under the leadership of then Captain Hyman G. Rickover. The function of the group was to measure neutron cross sections of various materials that might be of importance in the construction of propulsion reactors for submarines.

Clark Goodman was a truly remarkable man, although I have to confess that it took me some time to appreciate his unique talents. When I met him, he had just edited a book on nuclear propulsion for maritime purposes, and I had a strong feeling then that this would be an extremely important application of the new physics we were learning about the atomic nucleus. Clark Goodman had a broad-ranging curiosity about the world, and he had a really uncanny way of finding important practical applications for the basic physics that caught his interest. He also encouraged his graduate students to be independent, which was somewhat disconcerting in the beginning but turned out to be, when all was said and done, an extremely valuable experience.

There were four graduate students in the group, Robert M. Kiehn, Clyde L. McClelland, Janet B. Guernsey (who was a professor of physics at nearby Wellesley College), and myself. By far the most exciting event during those years was the accidental discovery of nuclear gamma rays produced by the newly discovered process of "Coulomb Excitation" by Clyde McClelland sometime in 1952. It turned out that the Coulomb Excitation process was particularly well suited for the study of certain low-lying nuclear energy levels that were important in verifying one of the important models of nuclear structure that was being developed at the time. For a while our little laboratory with its small electrostatic particle accelerator was a minor center of attention, and all of us basked in the limelight. It was also fortuitous for me at the time that Professor Victor F. Weisskopf took an interest in what we were doing. Some years earlier, Weisskopf had actually performed some calculations on Coulomb Excitation, and so we became, to some extent, his stepchildren. It was from Weisskopf that I learned during those years the importance of having a strong theoretical background in order to guide and inspire experimental work. Weisskopf has a true genius for developing simple yet comprehensive explanations of complicated phenomena, and I tried to absorb this ability from him as best I could.

During the years I spent in graduate school at MIT (1951–55), I almost completely forgot about space travel. I became totally immersed in the work in nuclear physics and became fascinated by the important potential applications. However, while I was at MIT an important event occurred that I hardly noticed at the time but that later turned out to have far-ranging consequences. This event was the publication by Collier's magazine during 1952 of a series of articles on space travel. The series was titled "Man Will Conquer Space Soon," and it was put together by a group of six or seven scientists, engineers, writers, and artists, the most prominent being Dr. Wernher von Braun of the U.S. Army's Ballistic Missile Agency and the leader of the German team that developed the V-2 rocket during World War II; Professor Joseph Kaplan of UCLA; Willy Ley, the popular science writer to whom I have already referred; and the artist, Chesley Bonestell. The series was edited by Cornelius Ryan, who later became famous for his books on several of the major battles of World War II.

This series of articles was a spectacular tour de force with some really breathtaking drawings and paintings. All of the ingredients of modern space operations were there in full color. There were reusable spaceships —or space taxis as Wernher von Braun called them at the time—and the whole thing was centered around the permanently orbiting space stations that would be used as bases for the myriad space operations that were contemplated by the authors. This series of articles was truly prophetic because the authors essentially outlined the program that we have executed since that time. It is remarkable indeed that more than thirty years ago there were people who had the imagination to lay out in detail what the possibilities were and how we could go about achieving them.

I completed my Ph.D. thesis at MIT in the summer of 1954 and stayed on for another year as acting head of the Neutron Physics Group while Clark Goodman was on a sabbatical leave. At the same time my wife finished her master's thesis at Boston University, and early in 1955 we decided to return to California.

Livermore

In 1955 I returned to Berkeley to spend a year working as one of the academic assistants of Edward Teller. Both my wife (who is a native-born Californian) and I wanted to return to California permanently, and this seemed like an excellent opportunity. I also wanted to learn something about theoretical physics, and I thought that the chance to work with Teller would further that objective in the best possible way.

There is no doubt that Edward Teller is one of the great physicists of this century. He is also a very great man. Teller had been a student in a class that my father taught in quantum mechanics at Karlsruhe University in Germany in 1927. There was, therefore, a family connection, and my father (once again) helped me to secure a position at the University of California, this time with his former student. In July 1955 I was appointed an assistant research physicist at the University of California at Berkeley. While I was sad about leaving MIT—I would especially miss some of the friends we had made during our years there—I did look forward very much to starting my new job in California.

My first year with Edward Teller was the most exhilarating intellectual experience I have ever had. I shared an office with Dr. Robert D. Lawson, and my job was to help Teller give his graduate course in advanced quantum mechanics. I also had the opportunity to work on some research with Lawson and one of Teller's graduate students, Dr. Auriol Ross Bonney. Teller had the same clarity of thought in physics that first impressed me about Victor Weisskopf at MIT. He also had a great appreciation of the practical applications of physics, and it was this facet of Teller's mind and imagination that I felt was most intriguing. Every morning there would be some new idea, whether it was digging a new canal across the isthmus of Panama with the help of nuclear explosives or putting a base on the moon. We had many vigorous debates about these things, and Teller always had strong opinions on any and all subjects. In spite of these vigorous arguments, Teller was never harsh or arrogant toward the people who worked with him. He was unfailing in his kindness and went out of his way to help people who were his associates. He inspired a degree of loyalty and affection among his assistants that was (and still is) truly unusual. The association with Teller that started in 1955 is still strong thirty years later. I continue to meet with him and listen to him whenever I have the opportunity to do so.

One of Teller's great passions was (and is) to apply the best science and technology possible toward the defense of the United States. He did much to reinforce my own feelings in that direction as well. Both Teller and I shared the experience of emigrating from Europe prior to World War II, so we understood instinctively the great dangers that can come from military weakness. Thus, when my year as Teller's academic assistant was complete, he had little trouble persuading me to join the research staff of the university's radiation laboratory's Livermore site (now the Lawrence Livermore National Laboratory).

The Livermore Laboratory was an exciting place in the early days of the effort to create nuclear weapons that would be of military value. When I joined the laboratory's staff in 1956, it was just four years old and there was a galaxy of bright young people then in their late twenties or early thirties who played leading roles in the technical developments. Among these were Drs. Harold Brown, John S. Foster, Jr., Herbert York (the director of the laboratory, who was himself barely over thirty years old at the time), Stirling Colgate, Theodore Merkle, and Michael M. May. All of these people and many others subsequently held very responsible posts and made critically important contributions to the national security. In addition, many of the senior people on the University of California's faculty in Berkeley, Professors Ernest Lawrence, Luis Álvarez, Edwin McMillan, and Glenn Seaborg (all of whom eventually won Nobel prizes) took an interest in the Livermore Laboratory and spent time with the younger members of the staff. I know that what I have just written may sound like name-dropping, but the easiest way to get some feeling for the electric atmosphere of the place at the time is to look at the quality of the people with whom we had the opportunity to work.

My own work was not related directly to nuclear weapons. There was at the time a large, high-intensity, but relatively low-energy particle accelerator at the Livermore Laboratory called the A-48. I was busy using this accelerator to make very precise determinations of the wave lengths of nuclear gamma rays that could be induced by the particle beam of the accelerator. (This work was done in collaboration with a very remarkable gentleman, Professor J. W. M. Du Mond of the California Institute of Technology. If Teller initiated me into the fraternity of theoretical physics, then Du Mond did the same for me in experimental work.) These data were useful to obtain very precise measurements of the positions of certain nuclear energy levels, a topic, at the time, of some interest.

None of this had anything to do with space, although there was considerable scientific interest in space activities at the time. The year 1957 was designated as the International Geophysical Year, and one of the projects to be carried out under the auspices of the IGY was the launching of an artificial earth satellite. This was project "Vanguard," and it was being carried out by a group of people who were then working at the U.S. Naval Research Laboratory. We knew that this work was going on, but it did not receive much public attention at the time.

In the late summer of 1957 (I think it was sometime in August), I met the person who would turn my thinking toward activities in space and would provide me with the first opportunity to make a small contribution to what would soon become a massive U.S. space program. This was Dr. Nicholas C. Christofilos. I had an office in one of the World War II "temporary" buildings that made up (and still make up) some of the office complex at the Livermore Laboratory. One morning I was told that Christofilos would be moving into our suite. This was interesting news because Christofilos was by then already recognized as an authentic genius. He had discovered, quite on his own while working as a young man in Greece during World War II, something called the "strong focusing" principle that turned out to be extremely important in the design of ultra-high-energy particle accelerators. As a result of this discovery, he was brought to this country and spent some years at the Brookhaven National Laboratory where he made important contributions to the design of the first "strong focusing" machine in the United States, the Alternating Gradient Synchrotron (AGS). When this work was completed, Christofilos developed a strong interest in nuclear fusion, and he also wanted to work on some classified ideas. Accordingly, he was brought to the Livermore Laboratory some time in 1956 or early 1957.

Since we were office neighbors, Christofilos and I quickly became friends. In due course he told me about one of the things he was working on, which had the code name "Project Argus." Christofilos had recognized that the geomagnetic field was configured like a magnetic mirror and that it would therefore be possible to trap energetic charged particles in the geomagnetic field if they were properly injected. He therefore conceived the notion of detonating a nuclear device in the geomagnetic field at the proper place so that the ions and electrons emitted by the device would be trapped in more or less stable orbits. Christofilos felt that if enough charged particles could be trapped in this manner, there might be some military value to the large radiation fluxes that would be created above the atmosphere. For example, the electronics of a nuclear armed reentry vehicle might be disabled by the radiation damage that the trapped charged particle "layer" would cause. Thus, the creation of such a layer of charged particles might be a good antiballistic missile weapon. Christofilos also thought that the synchrotron radiation emitted by the energetic trapped electrons might be useful in jamming the communication systems of potential opponents.

These were all fascinating ideas, and I offered to help Christofilos on a part-time basis. This offer was accepted, and so I started to work on the plan that Christofilos had outlined. The plan was to detonate a small nuclear device at an altitude of a hundred or so miles above the earth's surface and then to look for the trapped charged particles with radiation detectors mounted on sounding rockets. We would also deploy ships and ground stations to look for auroral effects and for radio waves caused by the trapped particles. It is remarkable that Christofilos had all of these ideas before the Van Allen belts, the naturally occurring trapped charged particles, were discovered by James Van Allen using *Explorer I* in 1958. I even remember late in 1957 hearing a prediction by Dr. Stirling Colgate that such naturally occurring trapped particles should exist. Since my technical expertise was in the field of radiation detection, I was asked by Christofilos to do some of the design calculations for the detectors that would be flown to test his ideas. I was working on a radiation background calculation one day when Colgate came into my office to pass the time of day. I showed him what I was doing, and he said: "You are underestimating the radiation background by only taking into account the scattered primary cosmic rays and reaction products that come back up through the atmosphere." He then went on to explain the mechanisms by which charged particles might be trapped. Specifically, he suggested that backscattered neutrons from cosmic ray events in the atmosphere could decay in the geomagnetic field, thus giving rise to energetic charged particles that would be trapped in the field. Unfortunately, I ignored Colgate's suggestion, and so I missed out on predicting what James Van Allen subsequently discovered!

Then, on October 4, 1957, an event occurred that was profoundly shocking to all of us interested in space and in technology development. This was the successful launch of *Sputnik I* by the Russians, who thereby became the first to place an artificial satellite into earth orbit. I vividly remember one night, standing on the roof of Building 157 at the Livermore Laboratory, watching the small bright dot of light move from one horizon to the other. I thought then that what I was watching would change our lives, and subsequent events have certainly proved that this was the case. *Sputnik* also had a profound effect on Project Argus. The plans for the experiment were much accelerated. Also, since there was now great pressure to put an American satellite into earth orbit, the people who were involved in this effort were brought into the planning for the Argus project. A group was established at Livermore under the general direction of Dr. Harold Brown, who was by then one of the associate directors of the laboratory. Professor James A. Van Allen of the University of Iowa and his graduate student, Carl McIlwain, were brought in as consultants, and it was decided that *Explorer IV*—which was then scheduled to be the fourth U.S. satellite —would carry the radiation detection equipment to look for the charged particles produced by the Argus detonations. Dr. George Bing, who was a member of the laboratory's theoretical division, and I were the principle representatives of the laboratory.

The Russian initiative had galvanized us into action, and once *Sputnik I* appeared there was no problem pushing the Argus project through. In due course the first American satellite, *Explorer I*, was launched on January 31, 1958. By that time the plans for the payload of *Explorer IV* were already far advanced. Perhaps the most interesting aspect of the *Explorer IV* payload is that it contained the first scintillation counter (a cesium iodide crystal detector) to be flown in space. It was designed to get good measurements of the charged particle energy distribution.

While I was working on the payload for *Explorer IV* in the spring of 1958, an amusing incident occurred that is worth repeating. In those days the American Physical Society held its major meeting in Washington, D.C., every May. I went to the meeting and attended the session at which James Van Allen announced his discovery of the now-famous Van Allen Radiation belts. I looked at the data and convinced myself that the phenomenon that Van Allen was reporting on was real. My first reaction was: "Oh my God, the Russians have blown up a large bomb without knowing what they were doing and have contaminated the geomagnetic field!" I quickly took a taxi over to the Pentagon to call on Dr. Herbert York, the former Livermore Laboratory director, who had just started his term as the Pentagon's first director of defense research and engineering. I told him about Van Allen's paper and about my worries about possible Russian nuclear explosions above the atmosphere. York cooled me down and told me that he was aware of Van Allen's results and that they were indeed radiation fluxes of natural origin. I still wish that I had listened more carefully to Stirling Colgate a few months before!

I continued to work on the Argus project in the spring of 1958, but another event occurred at the time that was a bitter blow to me and that caused me to leave the Livermore Laboratory. Sometime in early 1958 the laboratory reached the decision to shut down the A-48 accelerator. Since this machine was essential to the research work that my group (Robert Jopson, Edward Chupp, and Frank Gordon) was conducting on nuclear energy levels, this work had to be stopped. I tried to secure a faculty appointment on the Berkeley campus and failed. Accordingly, I accepted an offer to return to MIT as an assistant professor of physics starting in September 1958.

The Argus experiment was carried out in due course during August 1958. Two low-yield bombs (two kilotons) were detonated above the atmosphere. The bombs were placed on rockets that were launched from an aircraft carrier stationed in the South Atlantic Ocean. The charged particles produced by the detonations behaved according to predictions, and they were indeed trapped by the geomagnetic field. In 1959 there was a then-classified symposium to review the results of the experiment. I mention this only because it was at this symposium, held at Kirtland Air Force Base in New Mexico, that I first met a young Air Force major, Lew Allen, Jr., with whom I would have strong associations twenty years later.

Although it was great fun to return to MIT and work again with old friends, I was not happy there as a member of the faculty. The essential problem was that I had not really made up my mind whether I would pursue a purely academic career and spend my time doing purely fundamental research or whether I would go into more applied work and concentrate on things having more immediate engineering applications. What I tried to do at MIT was both, and that simply did not work—even in the short run. Professor Nathaniel H. Frank, who was then the head of the department of physics, had a number of conversations with me to try and help me make up my mind. I still remember his patience and kindness with much gratitude.

There was another individual who had much influence on my thinking during those years, and he was former Dean Morrough P. (Mike) O'Brien of the College of Engineering at the University of California at Berkeley. Mike O'Brien was spending a sabbatical year at MIT in 1959, and he spent a good deal of time talking with me. It was he who finally convinced me that my own talents and proclivities were not really in fundamental physics but rather in applied science, engineering, and management. It was Mike O'Brien who suggested that I should think about the possibility of joining the College of Engineering at the University of California in Berkeley and to combine a faculty appointment with work that I might want to do at Livermore. This sounded like a very interesting proposition, and I asked O'Brien to pursue it for me. Eventually, an arrangement was made where I would return to the Livermore Laboratory as P-Division leader and at the same time would hold an appointment as a lecturer in nuclear engineering at the Berkeley campus. I assumed these posts in the summer of 1960.

IV   The Space Race Starts

If *Sputnik I* was a shock, then Yuri Gagarin's first manned orbital flight in April 1961 was a disastrous political setback—at least this was the reaction of the newly elected president, John F. Kennedy. Shortly before his inauguration, the president-elect had received a report from a committee headed by Professor Jerome Wiesner of MIT on the subject of the American space program. This report contained a set of comprehensive recommendations about what should be done both in the military and the civilian parts of the program. The Wiesner report was generally very positive and aggressive, and it did recommend increased spending on scientific research, military payloads, and on space launch vehicles. On one point, however, the report was reticent, and that was Project Mercury, which was then under development by the National Aeronautics and Space Administration (NASA), the agency that was established late in 1958 to manage the nation's civilian space program. Wiesner and his colleagues pointed out that Project Mercury, which had the major objective of putting a man in earth orbit, was a risky proposition from a technical viewpoint and that the new administration would be blamed if there were any failures. Furthermore, the Wiesner committee argued that it was not clear that people in space had any real value in the sense that the scientific and military mission then contemplated could probably be carried out much more effectively and less expensively with unmanned spacecraft.

The arguments against Project Mercury made in the Wiesner report were swept away by the political impact of Gagarin's flight. The new president felt that he had to do something to respond to what the American public clearly perceived to be a challenge to this country's technological superiority. As it turned out, the newly appointed administrator of NASA, James E. Webb, had just the plan that Kennedy was looking for. For some years two separate groups, one at NASA's Langley Research Center and another at the U.S. Army's Ballistic Missile Agency at Huntsville, Alabama, had developed detailed and comprehensive plans for making a manned trip to the moon. Webb took these plans to the president and proposed that the proper response to Gagarin's flight would be for the United States to make the commitment to go to the moon. Kennedy was definitely interested in the idea, but there was much opposition to the plan among the president's senior advisors. Nevertheless, the president felt he had to do something, so he finally decided to go ahead. (A short and incisive insight into the process that led to the decision to go to the moon was provided by Hugh Sidey in a column that appeared in *Time* magazine on November 14, 1983. The whole issue of the magazine was intended as a memorial issue on the twentieth anniversary of President Kennedy's assassination.) Accordingly, on May 25, 1961, addressing a joint session of Congress, the president made his famous commitment "to put a man on the moon and return him safely to earth before this decade is out."

I was, of course, not involved in any of this at the time. Having returned to the Livermore Laboratory in the summer of 1960 as P-Division leader, I was busy trying to organize myself to assume my new responsibilities. At the time, P-Division had a total complement of about 150 people, and of these thirty or forty were professional physicists or engineers who had advanced degrees in their respective fields. The mission of the division was to perform basic research in nuclear physics that was intended to support the Livermore Laboratory's work in the design and development of nuclear weapons. This involved experimental work in nuclear cross-section measurements and nuclear reactions and also engineering support on occasion for certain nuclear weapons tests then being carried out at the U.S. Atomic Energy Commission's Nuclear Test Site. The division was well equipped for these tasks, having under its jurisdiction a large (90" diameter) variable energy cyclotron, a high-current electrostatic accelerator, and a small electron linear accelerator. Later on, the division also acquired a small nuclear research reactor and a linear electron accelerator.

In the waning months of 1960 I had to make some decisions about the research work that I would personally pursue and how this work would be related to the work of the division and to the laboratory's mission. Instead of returning to the earlier work on nuclear energy levels, my collaborators Robert M. Jopson and Charles D. Swift and I decided to strike out in new directions. We looked at some of the problems that people were having in the area of weapons design and decided that we would work on fundamental atomic physics rather than nuclear physics. This decision was based on the notion that nuclear weapons would become more specialized as time went on and therefore the designs would eventually depend as heavily on the atomic properties of the materials out of which the weapons were built as on the nuclear properties of the "fuel" being "burned" in the bomb. Since our technical expertise was in the field of radiation detectors, we started by building scintillation counters that could be used to look at relatively low-energy X-rays (quantum energies of the order of one to five kilo-electron-volts). Once we had such detectors, we would hunt for promising applications. The first of these was that we would make a comprehensive series of measurements of certain atomic fluorescence yields because these quantities turn out to be important in the calculation of radiation transport inside a nuclear explosion. Although I did not know it at the time, our initial interest in fluorescence yields turned into a very long-term scientific effort for me in the area of the physics of inner shells of atoms, which I still try to pursue in one way or another.

In the early 1960s P-Division became involved in space research in several different ways. Drs. Stewart D. Bloom and Lloyd Mann were building charged particle detectors that would eventually be flown on various NASA spacecraft. These experiments were intended to measure more precisely the distribution of particle species and their energy spectra in the Van Allen Radiation belts. This was essentially basic research, and as the division leader, I did whatever I could to encourage the work. My own research group was also drawn into space research, and this happened in two separate ways. One of the people I met in Berkeley was Professor Kinsey A. Anderson who was interested in cosmic rays and had done some notable work in that area. Anderson was at the time interested in looking at the passage of cosmic rays through the atmosphere, and he was engaged in flying payloads on high-altitude balloons to look at the secondary ionizing radiations produced by primary cosmic rays impinging on the atmosphere. One subset of these was low-energy X-rays, and our group made an agreement with

Anderson's group to provide them with the low-energy X-ray counters that we were building in our laboratory. This collaboration led to a large number of balloon flights, but, unfortunately, things turned out to be much more difficult than we had thought, so no concrete scientific results were achieved.

The second collaboration we embarked on turned out to be more interesting and productive. Ever since 1958 there was a "gentlemen's agreement" between Russia and the United States not to test nuclear explosives in the atmosphere. During the late 1950s each side had developed the technology to test nuclear weapons underground, and so more tests were conducted in that manner, thus reducing the radioactive debris scattered in the atmosphere. In 1961 the Russians broke the gentlemen's agreement and conducted a long and comprehensive series of nuclear explosions in the atmosphere. It was clear that we had to respond, and the entire scientific staff of the Livermore Laboratory was mobilized to participate in the 1962 nuclear test series that we were planning at that time. There were three tests in the series that were of particular interest to me; these were code-named Teak, Orange, and Starfish. These tests involved the detonation of high-yield (about one megaton) nuclear devices above the atmosphere, and, in due course, possibly because of my experience on Project Argus five years earlier, I was drawn into the planning process for these events.

I have already mentioned that one of the functions of P-Division was to provide technical support for nuclear weapons tests being carried out by the laboratory. One of the things we wanted to do was to observe the X-rays that would be emitted by the nuclear detonations above the atmosphere. Since we had some expertise in the art of detecting and measuring X-rays, we became consultants to the people in the laboratory's test division who had the responsibility to perform the diagnostics experiments. It was in this way that I met Dr. Frederick D. Seward, who was then in the high-altitude diagnostics group in the test division.

The test series was successfully completed in 1962, and the results of the Starfish shot confirmed what Christofilos had suspected back in 1957, namely, that it might be possible to damage electronic components of space and reentry vehicles with trapped charged particle belts or layers created by nuclear explosives. Several satellites then in operation were temporarily disabled by the trapped energy-charged particles produced by the bomb. At the same time, negotiations to prohibit the testing of nuclear devices in the atmosphere were initiated with the Russians. In 1963 a treaty—the so-called test ban treaty—was concluded that did indeed stop atmospheric nuclear testing.

When the Senate ratified the treaty, the late Senator Henry M. Jackson put a provision in the ratification act to maintain what was called the "Readiness Program." This program required that we maintain the readiness to test nuclear weapons in the atmosphere if it was deemed important to do so. The Readiness Program required periodic exercises of the diagnostic groups as well as the groups that actually conducted the nuclear detonations. Seward's diagnostic group was included in the program, and our own research group worked with him to launch small sounding rockets used to carry the X-ray detectors to practice the diagnostic operations. Since the rockets had to be fired anyway, we decided that we might as well do some physics with them. Dr. Riccardo Giacconi, then of the American Science and Engineering Corporation, and his collaborators had recently discovered that there were several stars that were strong X-ray sources. Accordingly, we modified our X-ray counters somewhat and began to look for stellar X-rays in each of the readiness operations we conducted. Our research on X-ray stars quickly paid off, and we began to get accurate measurements of the X-ray spectra emitted by a number of different stars. We worked on this program, launching several diagnostic rockets every year from the site at Barking Sands Air Force Station on Kauai Island in Hawaii. The project was terminated when the Readiness Program was abandoned in 1969. However, the X-ray astronomy program that we managed to conduct in those years kept my own interest in space operations alive and growing. The more I worked with the hardware, the more I became convinced that really important things would come from the efforts to go into space and to conduct experiments and operations there.

During this period we also managed to fly one of our X-ray detectors on a satellite. The satellite was an Air Force vehicle launched from Vandenberg Air Force Base in California. Our group at Livermore built the X-ray detector, received the data, analyzed them, and then published the results. The experiment, designed to measure X-ray emissions from the sun, turned out to be successful. All of this happened in 1966 and 1967, and the work provided many lessons for me that would prove to be most valuable in the coming years.

We also undertook another space-related program that was not as successful. As part of the Apollo program, NASA conducted two precursor programs to land unmanned spacecraft, *Ranger* and *Surveyor*, on the lunar surface. *Ranger* was a small spacecraft designed to make a hard landing on the lunar surface with the primary mission of taking a few pictures before it crashed. *Surveyor* was much more sophisticated in that it was designed to make a soft landing and then to conduct a number of experiments to make measurements of the surroundings. One set of these had to do with conducting an analysis of the lunar soil, and NASA decided to use nuclear methods to accomplish this objective. One of the P-Division physicists, Dr. Carleton D. Schrader, had the idea of using the inelastic scattering of neutrons to perform the analysis. Fourteen million electron volt neutrons would be produced by a small deuterium-tritium source located close to the lunar surface. These neutrons would be inelastically scattered by nuclei in the soil, and the nuclear gamma rays resulting from the scattering process would be measured by a scintillation detector. The spectrum of gamma rays observed in this way would be characteristic of the materials in the lunar soil.

We built a working mockup of the inelastic neutron-scattering experiment and convinced ourselves that the basic idea was sound—that is, using this method, one could indeed distinguish between the important types of lunar soils. We also started the procedure to persuade NASA to fly the experiment on one of the forthcoming *Surveyor* flights. Since *Surveyor* was a responsibility of the Jet Propulsion Laboratory, this put us in contact with Drs. Manfred Eimer, Alfred Metzger, and Jacob Trombka of the laboratory. Their responsibility was to evaluate our proposal compared to the other nuclear methods that were being proposed at the time. Unfortunately, it turned out that a simpler set of experiments combining neutron activation analysis and X-ray fluorescence measurements was much more effective in providing a good analysis of the lunar soil than the inelastic neutron-scattering method we had proposed. Consequently, nothing came of the effort to fly a payload on the *Surveyor* spacecraft.

Although I was not personally involved, there was another important program that the nuclear laboratories carried out during my tenure at Livermore that deserves discussion here. I have already mentioned that P. E. Cleator in his 1936 book hinted at the possibility of eventually using nuclear energy to propel spaceships, and that Willy Ley was very optimistic about this prospect in 1947. Once those controlled, self-sustained fission chain reactions were produced by Enrico Fermi and his group in 1942, it was quite inevitable that people would begin to think about using nuclear power to drive airplanes and spaceships. In due course, programs to achieve this objective were indeed initiated. In the late 1940s the Aircraft Nuclear Propulsion Program was started by the U.S. Air Force, and the Oak Ridge National Laboratory was given the responsibility to develop the nuclear reactor that would be appropriate for aircraft propulsion. Some years later, programs to create high-temperature nuclear reactors designed for rocket or ramjet propulsion were also initiated. After intense competition the rocket engine (code name "Rover") was assigned to the Los Alamos National Laboratory, and the ramjet (code name "Pluto") was assigned to the Livermore Laboratory.

The technical ideas that governed the Rover and Pluto programs were simple. A nuclear reactor would be built that would operate at very high temperatures—in excess of 3,000° F—and that would have tubes running through the reactor core through which the working fluid would pass and absorb energy from the hot core. In the case of the rocket engine (Rover), the working fluid would be hydrogen carried along in the spaceship. Hydrogen was selected because it would yield the highest specific impulse and therefore the most efficient propulsive system. The ramjet (Pluto) used air as the working fluid, and was designed to power vehicles designed to fly in the atmosphere. The air would enter the reactor through an appropriate inlet and would be exhausted after absorbing energy from the reactor as it passed through the core.

Working models of both the Rover and the Pluto reactors were built in the early 1960s. Both worked as designed, but it turned out that the promise that both Cleator and Ley thought they saw in nuclear energy for propulsion was not realized. The primary difficulty was that the potential gain of nuclear rocket or ramjet propulsion over that available by burning the appropriate fuels through chemical means simply did not justify the investment. In the case of a rocket the specific impulse depends on the temperature and inversely on the molecular weight of the exhaust products (specific impulse $\approx K \sqrt{\frac{T}{M}}$). The temperature at which rocket engines can be operated depends on the properties of the materials of which the rocket engine is constructed. As things turned out, the reactors could not be operated at temperatures that were any higher than the combustion chambers of conventional chemical rockets. The only gain of nuclear propulsion is that the working fluid (hydrogen) has a lower molecular weight than the combustion products that are produced in a conventional rocket (water and carbon dioxide). As things turned out, this gain in specific impulse was not enough to overcome some of the disadvantages inherent in using nuclear reactors, namely, relatively high weights and radioactive by-products. Accordingly, Rover was eventually abandoned as a practical way of obtaining a high-thrust booster.

When this story is considered, my decision in 1947 to study nuclear physics, in part because of the promise outlined in Ley's book for nuclear propulsion, was somewhat misguided. What happened was that the great progress made in high temperature materials technology and fluid mechanics was such that conventional rocket technology using chemical fuels remained consistently superior to any nuclear methods. This is still the case today.

In 1961 I was appointed associate professor of nuclear engineering at the University of California at Berkeley, and I continued on as the P-Division leader at the Livermore Laboratory. Three years later—in the summer of 1964—I gave up my administrative job at Livermore and became the chairman of the department of nuclear engineering at the university. My appointment as department chairman coincided with the start of the student rebellion in Berkeley. I was therefore somewhat distracted from scientific work for a while. I joined a faculty group (the Faculty Forum) that tried to maintain some kind of order on the campus, and eventually we succeeded. However, these were difficult years and I was not in a good position to contribute very much to the scientific work of my two collaborators at Livermore (Frederick Seward and Charles Swift). Nevertheless, I did whatever I could to maintain an interest in space activities of one kind or another. Aside from the work at Livermore, I befriended Professor Samuel Silver who was at the time the director of the University of California's Space Sciences Laboratory and who was well positioned to help me keep up my interest in space affairs.

In October or November 1968 I had a telephone message from Mr. James M. Beggs, who was at the time NASA's associate administrator for advanced research and technology based in Washington. When I returned the call, Beggs wanted to know if I would be interested in becoming the director of NASA's Ames Research Center in Mountain View, California. I knew nothing about the Ames Research Center at the time so I hedged and asked for time to make up my mind. I visited Ames and also went to Washington to talk to the relevant people at NASA headquarters. There were a number of factors that had to be considered and I vacillated for several weeks. At one point, I even told Beggs that I could not take the job but, fortunately, he asked me to postpone my decision. I was reluctant to leave the university and the work at Livermore; both were powerful magnets pulling me to stay where I was. On the other hand, to participate directly in the nation's space program by joining NASA was an unusual opportunity that I could not lightly reject.

Eventually, I did decide to move to Ames, largely on the advice of three very remarkable people, Messrs. C. W. (Bill) Harper, H. J. (Harvey) Allen, and Dr. Harry Goett. All of these people knew Ames well, having spent many years there, and in my conversations with them, they made an extremely persuasive case for going to Ames. I finally decided to take their advice and started work at Ames in February 1969. I have never regretted the decision.

V

The Start at Ames and the Space Shuttle

In retrospect, the move to Ames was a natural one for me. I had been engaged in space activities and in space science in a peripheral manner for a dozen years. Now I would join NASA and become a direct participant. One of the things that I learned while making the decision to go to Ames was the very direct political impact that space operations could have. My first visit to NASA headquarters to talk about going to Ames coincided with the *Apollo 8* flight in December 1968. I vividly remember standing in the anteroom of Jim Beggs's office in Federal Office Building 10B in Washington and listening to astronauts Frank Borman, Bill Anders, and Jim Lovell read the familiar passages from Genesis while their spacecraft was circling the moon. It was an awe-inspiring experience, and it undoubtedly influenced my decision to join NASA.

It was an exciting period. The Apollo program was just coming to a climax with the first manned lunar landing scheduled for July 1969. All efforts were oriented toward making it succeed. But even more important, many people were already beginning to think about what should be done after the lunar landing. Richard M. Nixon became the thirty-seventh president of the United States early in 1969, and in February 1969 he established the Space Task Group to develop what was then called the "post-Apollo" space program. The group was a very prestigious one, chaired by Vice President Spiro T. Agnew, and the members were Dr. Robert C. Seamans, secretary of the Air Force; Dr. Lee A. DuBridge, the president's science advisor; and Dr. Thomas O. Paine, the administrator of NASA. In addition, there were three "observers," Mr. U. Alexis Johnson, under secretary of state; Dr. Glenn T. Seaborg, chairman of the Atomic Energy Commission; and Mr. Robert P. Mayo, director of the Bureau of the Budget. Dr. Homer E. Newell, NASA associate administrator, was the principal NASA official working with the Space Task Group. In his letter establishing the group, the new president asked for a "definitive recommendation on the direction which the U.S. space program should take in the post-Apollo period" and requested an answer by September 1969. A staff was assembled in the office of the vice president to help write the report—*Apollo 8* astronaut Bill Anders was one of the leading members of this staff.

In addition to the Space Task Group, several other planning activities were initiated. One of these was an activity under the auspices of the president's science advisor, and the other was an in-house NASA group that was organized by Dr. George E. Mueller, who was then serving as NASA's associate administrator for manned space flight. Each NASA program office at the time had (and still has) a management council, which was chaired by the associate administrator who heads the program office, and the center directors reporting to the program office were the members. What George Mueller did was to expand the management council of the Office of Manned Space Flight (OMSF) to include all of the other center directors as well. It was in this way, as a brand-new NASA center director, that I became involved in the post-Apollo planning process.

Sometime in March or early April 1969, George Mueller held a meeting of his expanded management council at the NASA Electronics Research Center in Cambridge, Massachusetts. This meeting was an extremely important one for me because I would meet here, for the first time, the people who would lay the foundation of the U.S. space program for the next decade. There is no doubt that the dominant member of George Mueller's management council was Dr. Wernher von Braun, who was then director of the NASA George C. Marshall Space Flight Center. It was quite an experience for me to meet von Braun, the same man about whom I had read thirty years earlier in books on the early research on rocket propulsion. Von Braun was a most impressive character. He combined a sure-footed technical competence with a truly ebullient and optimistic personality. Von Braun had a passionate enthusiasm about space flight and space exploration that he communicated with great force and clarity. Another important figure on the OMSF management council was Dr. Robert R. Gilruth, who was then the director of NASA's Manned Spacecraft Center (now the Lyndon B. Johnson Space Center) in Houston. Von Braun's expertise was in propulsion, and Gilruth's in the area of spacecraft design. Gilruth was a very different personality from von Braun and was, in his own way, equally effective. If von Braun represented the visionary aspect of the space program, then Gilruth represented the deep technical care (or even conservatism) that was required to execute the visionary programs that von Braun and his friends conceived. Both Gilruth and von Braun became good friends, and they each had an important influence on my own thinking about the direction of the nation's space program. It was at this meeting of George Mueller's OMSF management council that I was introduced to the current concept of the space station and the reusable space vehicle, the space shuttle. Neither the space station nor the space shuttle were new ideas to the NASA planners. I have already mentioned the 1952 *Collier's* article in which von Braun and his coauthors first talked about space stations and space taxis. Later, in 1963, NASA conducted a very comprehensive study of a space station under the leadership of Charles J. Donlan. The ideas for the space station and the space shuttle were therefore very well developed, and the management council heard very comprehensive briefings of the plans already in existence.

It is most important to recognize that in the minds of the people who were making these plans the space station and the space shuttle were always part of the same program. In fact, the very name space shuttle was invented to describe the vehicle that would be used to "shuttle" back and forth to the space station. Ideas for the space shuttle were also well advanced as I was to learn a little bit later on. As the junior member of the group, I did not have much to contribute, but I listened carefully. In the scheme of things, Ames and the other NASA research centers (Langley and Lewis) played a supporting role by providing the technology needed to execute the larger programs that would be managed by the other NASA centers. I knew that whatever happened, Ames would have to become involved in supporting one or the other proposals that were being considered, and I wanted to make certain that Ames would have an important part to play.

The Mueller group also received briefings on the other studies of NASA's post-Apollo programs that were being carried out at the time. In the spring and summer of 1969 I remember hearing several status reports from the Space Task Group and also from the scientific panel established by the president's science advisor, headed by Dr. Lewis Branscomb, then director of the National Bureau of Standards and one of the founders of the Joint Institute for Laboratory Astrophysics. It was quite clear that a schism was beginning to develop between the ideas that were considered to be of the highest priority by the Mueller group and the briefings we were receiving from the others. In Mueller's management council the highest priority was always given to the creation of a new "capability" to operate in space, and what would actually be done with the "capability" was discussed but was clearly secondary in importance. In contrast, both the Space Task Group and Branscomb's committee concentrated on the mission objectives and put capability development as something that would have to be done in order to execute the missions that were being proposed. The division of thinking was quite natural because the Mueller group consisted of the people who headed the institutions that had the function of "capability" development. The other groups represented more broadly based constituencies—scientific, industrial, and political—and thus a much larger range of options was considered by them.

As the spring and summer of 1969 wore on, these ideas jelled into concrete draft reports. Roughly speaking, the principal recommendations being firmed up ran something like this (although there were considerable overlap and duplication among them):

Space Task Group. The primary recommendation of the Space Task Group was that the United States should make the commitment to undertake a manned mission to the planet Mars. The time scale on which this mission was to be executed was thirty years or "before the end of this century," as stated in the Space Task Group report. There were a great many other recommendations as well, including both the space station and a "chemically fueled" space shuttle, but these were subsidiary to the primary goal. In addition, there was a melange of scientific missions as well as fairly detailed consideration of the importance of NASA's work to the Department of Defense.

Branscomb committee. The Branscomb committee issued a very comprehensive report in March 1970. It makes no strong recommendations, but it clearly questions the recommended goal of the Space Task Group by pointing out that the predictable advance in automation and the technology of unmanned space flight might make the Space

Task Group's goal untenable. The Branscomb committee recognized the importance of Mars as the next step and put it at high priority. But, the committee implicitly questioned the necessity of a manned mission to Mars. The report also contained a large number of other important statements, particularly those dealing with the conduct of astronomical observations from space. They put very high value on the creation of orbiting astronomical observatories and on the necessary facilities to support this type of work.

Mueller management council. The viewpoint of Mueller's management council was clear and simple. The next step was to build a space station and to develop the reusable shuttle vehicle to service it. This view was clearly and concisely expressed in the program content of NASA's report to the Space Task Group in September 1969. The report stressed the importance of "reusability" (that is, the shuttle) and of developing a space station to serve as the essential building block for more ambitious missions in the future. Once again, the report also contained numerous other things, but the central focus of the recommended program plan was the space station and the space shuttle. In short, George Mueller had won the debate inside NASA to concentrate on the development of "capability."

Needless to say, I was completely enthralled by the opportunity to participate in all of these activities. The ideas that were being considered had an intrinsic fascination for me that was clearly irresistible. The problem I had was that I also believed that the optimistic assessment of future NASA budgets inherent in all three camps (STG, Branscomb's, and Mueller's) were unrealistic. I had just left the Berkeley campus where I saw the political passions that were unleashed by the Vietnam War and by President Lyndon Johnson's ambitious attempt to rectify, by government fiat, the racial and economic injustices that continue to bedevil the nation to this day. I simply did not believe that any of the ambitious plans to which I was exposed in the summer of 1969 could actually be executed. I was still a newcomer to the government planning process, and I did not realize that all of these plans were "unconstrained" and that they would be subject eventually to the normal budget processes. There were two other important things that I learned as a result of the discussions of 1969. One had to do with the perennial question of manned versus unmanned space flight, an issue that would continue to bedevil our efforts to push the space station program through the administration and the Congress fifteen years later. The second was that the people who tended to win the arguments were those who had specific and well-focused plans, whereas the losers tended to be those who tried to accommodate all viewpoints in one grand design. As it turned out, these lessons were to stand me in good stead later on.

I cannot tell this story without mentioning the Apollo program and especially the landing of the *Apollo 11* astronauts on the moon. It is perhaps best to begin by asking why post-Apollo program studies were necessary at all? Did not the Apollo program itself contain the ideas for a logical succession of programs that would follow? Unfortunately, this was not the case, and the reason had to do with a decision that was made in the summer of 1962, a year after President Kennedy initiated the program. The problem can be stated as follows: it was clear that a direct journey to the moon, that is, one which was not dependent on some kind of an intermediate staging base, would be prohibitively expensive. Therefore, staging would be required, and the question was how this could be accomplished. There were two essential choices: one was to do the staging from earth orbit (the method of earth orbit rendezvous), and the other was to stage from lunar orbit (lunar orbit rendezvous).

Two NASA committees, one chaired by Dr. George M. Low and the other by Dr. Bruce T. Lundin, both recommended that the earth orbit rendezvous method be adopted. It was deemed to be the lower-risk option because the necessary techniques would be practiced in the forthcoming Gemini flights. Furthermore, and this was something that was even then (in 1962) close to the thinking of the people involved in the planning process, the earth orbit rendezvous method led automatically to the creation of an earth-orbiting space station of some kind. There would, therefore, be a legacy from the Apollo program that would permit us to conduct other missions using the space station as a staging base. Unfortunately, this is not how things worked. Dr. John Houbolt and his colleagues of the NASA Langley Research Center in Hampton, Virginia, demonstrated beyond a shadow of doubt that the lunar orbit rendezvous method they were advocating would be less expensive. More to the point, Houbolt convinced the NASA management that the only way to meet President Kennedy's schedule of "placing a man on the moon and returning him safely to earth before this decade is out" would be to use the method of lunar orbit rendezvous. Houbolt was right, of course, and his proposal was finally adopted. President Kennedy's objective was duly accomplished, but we paid a price: the Apollo program had no logical legacy.

In spite of all of this, there is no question that the landing of Neil Armstrong and Edwin (Buzz) Aldrin on the moon on July 20, 1969, was an epochal event. I had the good fortune to be in the Mission Control Center at the Manned Spacecraft Center (now the Johnson Space Center) on the day of the landing, and it was clearly one of the high moments in my life. My wife and I had been invited to Houston for the event because, as a newly minted NASA center director, I rated such an invitation. As it turned out, the director of science and applications at the Manned Spacecraft Center was Dr. Wilmot N. Hess, who was an old friend from Livermore days. There were also new friends. Dr. Alfred J. Eggers, who was at the time one of NASA's assistant administrators, had spent many years at Ames and became one of my guides in understanding the intricacies of NASA politics. There was also Dr. Harrison (Jack) Schmitt, who was then a scientist astronaut, who would go to the moon on *Apollo 17*, and who would later serve in the U.S. Senate. All of these people were important in their own right, and Eggers and Schmitt would have particular influence on me in years to come.

The climax was, of course, Armstrong's first step on the moon. As it turned out, the actual landing was in the evening. Armstrong and Aldrin then slept for a while, and in the early morning hours (Houston time) we sat in the viewing room behind the Mission Control Center (MCC) watching the flickering television repeater to see Armstrong's historic first step on the lunar surface. After the first step, I remember staying on for an hour or two to watch some of the other things that might occur. I stepped out of the viewing room for a short time and went down the half flight of stairs to the hall where I ran into Bob Gilruth. He was exultant but also worried. "Will we get them back?" he kept asking of no one in particular. It was from Gilruth that I learned the terrible tension that went with sending people into this new medium of space. The few hours I spent at the MCC on that day were not to be equaled until a dozen years later when I watched the space shuttle *Columbia*'s first takeoff from the same place.

Shortly after the *Apollo 11* mission was completed, President Nixon hosted a large formal dinner to celebrate the event. The dinner was scheduled for August 13, 1969, and was to be held at the Century Plaza Hotel in Los Angeles. As a NASA center director, I was invited to attend the party along with my wife. I want to describe what happened at the dinner, not because it was intrinsically important, but because it illustrated the attitudes toward the space program that were prevalent at the time. My wife and I were pleased to be able to attend, and we were excited because this was the first time that we had been invited to go to a party hosted by the president of the United States. Accordingly, we made the necessary preparations to attend.

On the appointed day, we flew to Los Angeles and drove to our hotel. We changed clothes and then drove to the Century Plaza. In the politically heated atmosphere of the time, there were almost certain to be pickets in front of the Century Plaza, and there was the possibility of a more active demonstration as well. Sure enough, the pickets and the demonstrators were out in force. A major feature of the demonstration was a huge sign with the legend "Fuck Mars" printed on it in large letters that the demonstrators had somehow been able to hang along the upper floors of one of the office buildings across the street from the Century Plaza. The same message was clearly repeated on signs that some of the demonstrators carried. It was very apparent to us where the demonstrators stood on the value of the space program and on some of the plans then being considered for the post-Apollo effort. (I thought that the way the message was presented was also typical of the intellectual level on which the protests of the 1960s—with few exceptions —were conducted.)

We picked our way through the crowd of people in front of the hotel and finally managed to get into the ballroom. Needless to say, the atmosphere was rather different once we were inside. We sat at the same table with the actress Diane Baker, the comedian Jonathan Winters, and the then secretary of health, education and welfare, Robert Finch, and his wife, and the famous aviatrix, Jacqueline Cochran. The atmosphere was festive and patriotic and the president made a rousing speech before introducing the three heroes of the occasion, Neil Armstrong, Buzz Aldrin, and Mike Collins. After the formal part of the program was completed, the president walked around the ballroom and greeted some of the guests. I remember him coming over to our table and spending a few minutes speaking with us. He was

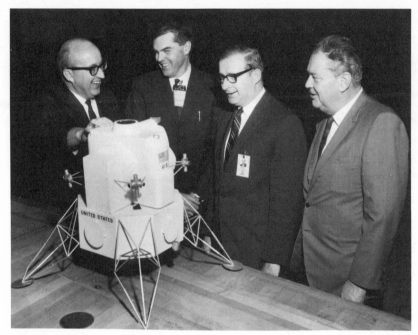

Clarence A. Syvertson, Mark, Thomas Paine, and H. Julian Allen at Mark's
inauguration as director of Ames. Paine was soon to leave NASA.

in a very euphoric mood, and I wondered whether he was at all concerned about the exhibition of the deep divisions that then existed in
American society we had just witnessed before the dinner. I suppose
that he really was concerned because, in his own way, he tried to do
something about them. Unfortunately, it was not enough and these
divisions, together with his own character flaws, would eventually destroy
his presidency.

The Space Task Group report that President Nixon had requested
early in 1969 was submitted in September of that year. Even before it
was finished, it was quite clear to me that the political climate in the
country was such that the ambitious program outlined in the report
would not be executed. When it became apparent that the Space Task
Group's recommendations were stillborn, Tom Paine decided to leave
NASA. I remember Homer Newell informing us of Paine's decision at a
meeting at Woods Hole at which we were discussing future plans and
prospects early in 1970. Fortunately, Paine's deputy, Dr. George M.

Low, turned out to be a man of really unusual talent, tenacity, and strength of character. Even though Low never became administrator, he provided the critical leadership in 1970 when it was necessary to regroup from the setback of the rejection of the Space Task Group program. It was Low who injected the appropriate degree of realism into the deliberations and who saw to it that the best recommendations were actively pursued by in-house NASA study groups. This turned out to be crucial in developing the post-Apollo program when the time came to make some real choices.

In the autumn of 1969 Gilruth and his brilliant associate, Dr. Maxime Faget, visited Ames to talk about the space shuttle. Faget had been responsible for the engineering design of the Mercury spacecraft and had also made major contributions to Gemini and to the Apollo program. In addition to being one of the nation's most brilliant engineers, Faget is also one of the most persuasive. Faget delivered a lecture on the space shuttle to the Ames senior engineering staff in which he eloquently explained the possibilities, the difficulties, and the solutions as he saw them. Faget is a man with an impish sense of humor, so he wound up his lecture by flying some small balsa wood models of space shuttles that he had brought with him around the room. "See, it flies!" he kept saying with a wide grin on his face. In spite of the skeptical attitude that I brought to all my evaluations of NASA's plans, I was impressed by what Gilruth and Faget had to say about the shuttle. They were clearly people who had to be taken seriously. I did not know it at the time but this visit was the beginning of a long and fruitful collaboration between Ames and the Johnson Space Center on the shuttle program.

At about the same time and into 1970, a shakeout of the planning process in NASA headquarters was in progress. As a first step, Wernher von Braun was persuaded to leave his post at the Marshall Space Flight Center to come to NASA headquarters to supervise the development of NASA's post-Apollo plan. This was in late 1970 and von Braun held the title of deputy associate administrator—which was not really descriptive but it served well enough. I came to know him well during the short time he spent with NASA in Washington and I had the opportunity to see his mind at work. As the more realistic program plans were developed during the tenure of George Low as acting administrator, the space shuttle became more and more important relative to the other elements of the program being considered. George Mueller's plan to put the space station and the space shuttle at the highest priority level had been the most compelling, and it was, at least implicitly, adopted by my senior colleagues. However, even this plan was too expensive for the political environment as it existed at the time, and it was apparent that at some point a decision would have to be made as to whether to do the space station or the space shuttle first. With the funds likely to be available, it simply would not be possible to do both of these things in parallel.

I have to confess that I was not an enthusiastic supporter of the space shuttle program at the time. My own background in space science had conditioned me to think in terms of unmanned space operations and I felt that the presence of people might actually complicate things. (This is a view still held today by many active space scientists.) I also felt that the space shuttle then being considered was too large and that it would be best to evolve a space shuttle from a series of smaller vehicles much along the same lines as the Air Force's recently canceled Dyna-Soar Program. (Dyna-Soar was canceled by Defense Secretary Robert McNamara on December 10, 1961.) I remember at this time having a number of conversations with Dr. George Morgenthaler of the Martin Company about the possibility of modifying Martin's X-24B to ride on top of a Titan and to become a reusable orbiting spaceship. The X-24B was much smaller than the proposed shuttle vehicle, and I felt that it would therefore be a more appropriate first step.

It was von Braun as much as anyone who persuaded me that the space shuttle as proposed by the Office of Manned Space Flight (OMSF) was indeed the right first step in realizing the space station plus space shuttle plan. I had a number of in-depth conversations with him in 1970 and 1971 during which he explained his rationale. First, he made a very persuasive argument that people were absolutely essential if we were going to conduct really sophisticated space operations. He felt that there was really no substitute for the presence of human judgment and imagination on the spot and that these qualities would be important in any advanced mission. People can take advantage of unexpected opportunities and deal with emergencies. These arguments were, of course, similar to the ones that people in the Air Force use to justify manned combat aircraft for most important purposes. I found these arguments to be persuasive, and they went a long way toward making me a supporter of the space shuttle program. There was another important matter over which von Braun apparently had considerable influence. In the debate over the sequence between the space station and the space shuttle programs, von Braun was strongly in favor of doing the space shuttle first. He felt that the establishment of a space station without something like a large space shuttle made no sense. He felt that a really effective space station would have to be assembled on orbit, and this would be impossible to do effectively with expendable launch vehicles. He also made two other points:

1. The space shuttle is the technically more difficult part of the whole program—that is—it would be harder to build the shuttle than the space station. Thus, the pace at which the shuttle program can be executed would eventually determine the time at which a space station could be deployed.

2. Once the space shuttle was built, the operations with the shuttle would attract considerable public attention, and this, in turn, would make it easier to persuade the political system to commit to a space station program. On this last point, as it turned out, von Braun was absolutely correct. There is no doubt that shuttle operations and the public attention generated was a decisive factor in the space station decision.

During 1970 all of the discussions of the space shuttle revolved around a fully reusable system. When Dr. James C. Fletcher, who was then president of the University of Utah, was appointed administrator of NASA in early 1971, negotiations with the White House to fund the post-Apollo program were resumed. Fletcher, as the new administrator, had to establish himself and advocate a program that would carry NASA through what clearly would be some very lean years. Fletcher is an extremely intelligent person with a very astute sense for the political atmosphere. I felt very comfortable with Fletcher as the NASA administrator because I thought that he would read the political situation correctly and that he would try to develop a program for NASA that could be sustained even under very difficult financial circumstances. At the time of Fletcher's appointment, the country was distracted by internal disorders brought about by opposition to the Vietnam War and Fletcher judged, correctly, I believe, that the ambitious plans proposed by the Space Task Group could simply not be sustained. Fletcher also recognized the importance of carrying out what is called a "balanced" program in which care is taken to make certain that each of NASA's major constituency groups take part in the program.

What evolved as a result of Fletcher's early planning was a NASA program that had two major elements: First, a "reusable" space shuttle but one that was cut back from the completely reusable vehicle originally proposed by George Mueller's committee. The reusable booster was discarded for a large external fuel tank for the shuttle main engine and two large solid fueled boosters necessary to get the whole thing off the ground. Second, a program would be initiated to put a soft-landing spacecraft on the planet Mars. (This became Project Viking.) In addition, there would be an active but rather modest program of smaller things that would serve to provide the appropriate balance that Fletcher thought was so important. Thus, what survived from the imposing plans made a year and a half earlier was the priority put on the planet Mars by the Space Task Group (Viking) and George Mueller's reusable spaceship.

Early in 1970 Mr. Dale Myers succeeded George Mueller as associate administrator for Manned Space Flight. It fell to Myers to lead the work of redefining the shuttle configuration and to make certain that the preliminary design and engineering work were properly executed. Myers, who is an experienced and exceedingly competent aerospace executive, performed these tasks with great distinction.

I have already said that I was not really an enthusiastic supporter of the space shuttle program when it was started—I was, in fact, something of an inside critic. Two things were significant in eventually changing my mind: one was the *Apollo 13* accident and the other was that I had further conversations with Wernher von Braun. It is worth discussing each of these briefly.

The *Apollo 13* flight was launched on April 11, 1970, and trouble developed about fifty-six hours after launch. A day later, George Low called and asked me to come to Houston to sit on the Accident Investigation Board. This was my first real introduction to the technology of manned space flight. The chairman of the board was Dr. Edgar M. Cortright who was then the director of the Langley Research Center and who had served earlier as a deputy in George Mueller's OMSF. He was (and is) a first-class engineer who thoroughly understood the essential technical problems. Dr. John Clark, then the director of the Goddard Space Flight Center, was also a member of the board and his expertise in deciphering telemetry records turned out to be the critical element in working out what actually happened. Finally, there was Neil Armstrong, who was then the most experienced of the astronauts and from whom I absorbed my first sample of the lore of space flight.

The *Apollo 13* mission was the third flight to the moon and was commanded by Jim Lovell, who was at the time our most experienced command astronaut. The other members of the crew were Jack Swigert (lunar module pilot) and Fred Haize (command module pilot). The investigation board quickly established that there had been an explosion in one of the oxygen storage tanks in the command module. The explosion was relatively benign so that the whole spacecraft was not lost. However, enough damage was done to the command and service modules so that the whole mission had to be aborted. A series of complex work-around steps were developed to substitute the lunar module engine, power systems, and life support systems for those in the damaged command and service module. This work was done by the astronauts and the ground controllers working closely together through the existing communications links. The landing on the moon was aborted and eventually the spacecraft and crew returned safely after a difficult return flight.

My service on the *Apollo 13* Accident Board served to convince me that manned space flight was not only important but was in fact central to the whole "enterprise in space." The dramatic return flight was ample demonstration of the critical importance of human judgment and intelligence on the spot. I could easily extrapolate what I learned to other situations where the objective was not to save a mission but to perform some other complex function for which automated systems or systems controlled from the ground were designed. The conviction that human beings would be sufficiently valuable in space to justify the expense of sending them there slowly began to grow in my mind starting with the *Apollo 13* experience.

The association with von Braun was not as dramatic as *Apollo 13* but was, on balance, even more important in erasing my initial biases against manned space flight. Von Braun was an enthusiastic sailor, a passion that I shared with him. Every time he visited Ames during the early 1970s, I made it a point to arrange a sailing trip for him. I vividly remember one day in late 1972 or perhaps early 1973 when we took him on one of the sailing races that we had in the winter time on San

Mark and Wernher von Braun on San Francisco Bay in late 1972.

Francisco Bay. The last Apollo flight (*Apollo 17*) had just been finished and one of the young members of our racing crew asked von Braun when we would ever return to the moon. As it happened, we were on a long downwind leg with not much to do so von Braun had a chance to hold forth on his views. He drew the analogy between the exploration of the moon and the exploration of Antarctica. He pointed out that after the epoch-making trips of Scott and Amundsen to the South Pole in 1911, not much of interest happened until an "enabling technology" as he called it, was put in place. In the case of Antarctica, this enabling move was the airplane, which was first used for Antarctic exploration by Richard Evelyn Byrd and Lincoln Ellsworth in 1929. Von Braun asserted that the "enabling technology" in the case of the moon would be the space station. In making this claim, von Braun was, of course, simply restating the arguments that Von Pirquet first made in his seminal articles of 1927 and 1928 which had to do with the fact that once you were in earth orbit you were three quarters of the way to wherever else you wanted to go. The whole crew of our boat found von Braun's lecture fascinating to say the least. This is one example of the nature of our discourse, and by the beginning of 1972, I had become a convert.

In the meantime—that is, in the summer and fall of 1971—Jim Fletcher took charge of NASA. He had persuaded the administration to approve the space shuttle program as the centerpiece of the nation's space effort. President Nixon signed off in January 1972, and we were on our way. There is no doubt that this was a very significant success for Fletcher since there was, in fact, some formidable opposition to Fletcher's plan. Before President Nixon approved the proposal to build a space shuttle, he asked his science advisor, Dr. Edward David (who had replaced Lee DuBridge in 1970) for an opinion. David established a committee to look into the problem, and the committee produced what was, in essence, a negative report. The central argument in opposition to the space shuttle proposal was that it was not "necessary" to put people in space. All of the objectives of our space program could be achieved more easily and at far less expense by using unmanned spacecraft—so the assertion went. This argument continues to be widely used and believed. I do not know whether it is right or not because I do not know how to quantify the value of putting people in space. What I do know is this: The argument is almost certainly beside the point. People will go into space for reasons that have nothing to do with the cost-benefit analysis of the type advocated by David's committee. President Nixon understood this point and ignored the advice of David's group in approving the shuttle program. President Nixon understood the political value of putting people in space, and this is ultimately why he did not listen to his scientific advisors. This argument is still going on and it has been a recurrent one in the entire history of the U.S. space effort. (Remember that President Kennedy made the decision to go to the moon over the opposition of his scientific advisors as well.)

On January 5, 1972, Jim Fletcher met with President Nixon in San Clemente and secured the final approval of the space shuttle program. Now, the fat was in the fire and we had to produce. One thing that happened was that each of the NASA centers was assigned a significant role in the development of the space shuttle. Ames drew the primary role in helping to develop the thermal protection system and also some supporting work in the areas of guidance and control. In February 1972 George Low called a meeting of all the center directors to discuss the readiness of the various technologies involved and the risks that were entailed in the creation of the space shuttle. Unfortunately, I could not attend the meeting, but I did write a letter to my immediate supervisor, Mr. Roy P. Jackson, the associate administrator for aeronautics and space technology, on February 15, 1972. The letter is reproduced in its entirety in appendix 1. Our assessment was that we were ready to go, but that the technology to build a reusable thermal protection system was not in place for the projected schedule of shuttle development. Fortunately, there were delays in the program so the reusable system was created. It is an interesting sidelight that in our letter (and it was not written by me alone, but rather by all of the Ames people who were eventually to become involved in the shuttle program) we cautioned that one of the problems we were facing was how to glue the thermal protection tiles to the structure of the shuttle so that they would not come off. Our advice was not heeded and much trouble would come as a result—but I am now getting ahead of the story.

VI

The Development of the Shuttle, Colonies in Space, and Politics

The commitment to the construction of the space shuttle that Jim Fletcher had secured from the Nixon administration was a major milestone for NASA and for the American space effort. Even though we were convinced that the job could be done, it turned out to be more difficult than we anticipated in 1972. We promised to build a reusable space ship, we promised to fly it for the first time in 1978, and we promised to do all of this for a development cost of the order of $5.5 billion in 1972 dollars. How well we fulfilled these promises will be discussed later on. Whatever success we did have in the end was due in large measure to the remarkable and hard-driving executive who eventually succeeded Dale Myers as NASA associate administrator for manned space flight, Mr. John F. Yardley of the McDonnell Douglas Corporation. If Mueller was the one who provided the initial push for the project in NASA, and Myers led the consolidation of the project, it was Yardley's engineering and management talent that made the ultimately successful execution of the project possible.

During the early 1970s, following President Nixon's commitment to build the shuttle, a substantial effort was mobilized at Ames to support the work. Most of this effort was devoted to the development of the reusable thermal protection system of the shuttle and to the development of the proper aerodynamic configuration using the extensive wind tunnel facilities existing at Ames. (The first effort was led by Glen Goodwin, Dean Chapman, Vic Peterson, Howard Larson, and Howard Goldstein, and the second by Leonard Roberts, Richard Petersen, Byron Swenson, and Harry Hornby.) These efforts turned out to be of critical importance, and there is no doubt that the shuttle program would have failed without the intensive supporting effort carried out not only at Ames but at the other NASA centers as well.

Early in the shuttle program, a decision was made to name the Johnson Space Center in Houston as the "lead center" for the development of the space shuttle. This meant that the responsibility to manage the project in detail was given to the Johnson Center, subject to the overall direction of Yardley's program office located at NASA headquarters. For those of us who were in the business of supporting the shuttle program this meant that we had to establish close relationships with the people at Johnson in order to perform our functions. I have already mentioned the remarkable Robert Gilruth who was at that time the director of the center. When I first met Gilruth he seemed to me to be a counterweight to Wernher von Braun's brilliant and visionary ideas. I soon found that he was much more than that. First, he was a visionary in his own right—only a very quiet and effective one. It was Gilruth who led the team at the Langley Research Center that, during the 1950s, developed the plans for going to the moon that subsequently fulfilled President Kennedy's mandate. Second, he established and led the Johnson Space Center and gathered around himself a number of unusually able and brilliant people including Max Faget, Jim Elms, George Low, Chris Kraft, and many others. Finally, it was Gilruth who, more than anyone else, established the principal ground rules for how we operate in space. Fortunately for me, he took a liking to me and we have enjoyed a close and productive friendship for well over a decade.

In 1970 and 1971 Gilruth invited Ames to participate closely with his people on the shuttle program, and this started a long series of visits that I would make to Houston during my service at Ames until I left in 1977. We worked with the Orbiter Project Office led by Aaron Cohen and his deputy, Milton Silveira. Both became good friends and strong supporters of the work being done at Ames to make the shuttle a reality. My early skepticism regarding the technical feasibility of the shuttle slowly faded, and by 1972 I was completely convinced that the program could be successfully executed, and this was due, in large measure, to the respect I had for Cohen, Silveira, and their extremely talented coworkers. The Ames group continued to work with the Johnson Orbiter Project Office after I left Ames up to the very end of the shuttle development program.

I was not as happy with the economic and political justifications that we were using for the space shuttle as I was with the technical progress being made. Since the political support for the space program was at a low ebb in the early 1970s, Fletcher and his staff were driven to make economic justifications—something that was new to NASA since these were not necessary during the Apollo program. In due course, a consulting firm, Mathematica, was hired to produce a thorough economic analysis of how the shuttle would operate. Mathematica was headed by the distinguished Professor Oskar Morgenstern and he designated Dr. Klaus P. Heiss as the project leader for the NASA work. Fletcher asked Heiss to calculate under what circumstances the shuttle could be operated less expensively than conventional launch vehicles. Heiss answered this question with a simple argument. If you take the cost of the consumables (i.e., fuel, etc.) as being equal for the shuttle and for conventional launch vehicles, then conventional launch vehicles are cheaper per unit weight in orbit for low launch rates because they do not include the complexities required for sustaining people in space; at high launch rates, the cost per unit weight in orbit of the shuttle must be less because less is thrown away. Heiss reasoned, therefore, that there is a flight rate at which there is a crossover between the cost per unit weight in orbit of the shuttle and of a conventional launch vehicle. He calculated that the number of flights at which this crossover occurred would be roughly thirty flights per year. However, the NASA management was more optimistic about the rate of future space activities, and a "mission model" of about fifty flights per year was eventually adopted.

The Mathematica study was completed in January 1972, and Heiss visited all of the NASA centers to present his results. I remember that we had a lively discussion with Heiss during his visit because the flight rate that both he and NASA headquarters were projecting implied an enormous budget for space activities to pay for all of the payloads that would be flown. Both thirty flights or fifty flights per year were substantially higher than the launch rate in the early 1970s. (The actual number of launches that the United States made in 1973 was 23.) Many of us were unhappy with the conclusions because we could not honestly reconcile ourselves to the shuttle launch rates that were being forecast. It is ironic that when Jim Beggs and I became administrator and deputy administrator, respectively, in 1981, almost the first thing we did was to change the projected launch rate from fifty per year to twenty-four per year in our budget estimates. In my judgment, this number was still a little bit too high. The whole question of shuttle costs still remains to be resolved, but I will postpone a discussion of this until the concluding chapter.

In spite of the fact that Apollo was essentially a dead end from a technical viewpoint, two important programs grew from the Apollo effort that profoundly affected my own thinking about the space station. One was Skylab and the other was the Apollo-Soyuz mission. As it turned out, Ames had very little to do with either one of these programs so that I can speak of them only as a spectator. Skylab was important for two reasons. One was that the failure to deploy the sun shield and the solar panel after the launch forced an impromptu demonstration of how important people were to operations in space. The installation of the "parasol" sun shield and the deployment of the solar panels in their "space walk" by Pete Conrad and Paul Weitz were the clearest kinds of demonstrations one could possibly imagine. In short, they saved the mission. The second reason was less dramatic but perhaps more important in the longer run. Skylab was really a prototype space station, and the whole project was therefore a precursor as well as a real demonstration of what could be done with a permanently manned space station. It is one of the real "might have beens" of the U.S. space program to contemplate what would have happened if the shuttle had been completed on the schedule contemplated in 1972 and if Skylab could have been rescued from the ultimate fate that overtook it in 1979.

The Apollo-Soyuz program did not have any far-reaching technical consequences, but it demonstrated the political and symbolic importance of space operations, and this was ultimately also something that would be important in the effort to persuade the Reagan administration to go ahead with the space station. The "handshake in space" that was the real objective of the Apollo-Soyuz program was conceived by Henry Kissinger to symbolize the foreign policy of "détente" that he was pursuing with the Russians in the early 1970s. George Low, who was acting administrator at the time, accepted Kissinger's proposition and took overall charge of the program. Low did a brilliant job and, as far as one can ever be sure of subjective things of this kind, Kissinger's objectives were realized. Both Skylab and Apollo-Soyuz were executed with leftover hardware from the Apollo program and, therefore, were one-shot propositions. From my viewpoint, the most important legacy of Apollo-Soyuz was my meeting with then Brigadier General Thomas Stafford who was the U.S. mission commander. He was to become a good friend and an important collaborator later on. I met Stafford in late 1975 after the Apollo-Soyuz flight when he brought his two Russian colleagues to visit Ames. The Russians (Aleksey A. Leonov and Valery N. Kubasov) were both very intelligent and highly trained professionals. Their visit illustrated to me at firsthand the Russian commitment to manned space flight and their technical ability to carry out such missions. They are good, but not, of course, as good as we are.

Early in the fall of 1974, I had a visitor who would play an important role in developing and guiding the nation's thinking about space activities and who would revive my own interest in space stations. This was Professor Gerard K. O'Neill of Princeton University. I had known O'Neill slightly in the 1960s when he attracted considerable attention by advocating and then developing the first "colliding beam" particle accelerators. The colliding beam principle has turned out to be extremely important in modern high-energy physics, and it is the fundamental contribution that makes possible the study of particle interactions at ultra-high energies. O'Neill did not want to talk about physics but about space stations. In our preoccupation with the development of the space shuttle and with the management of the other programs at Ames, I had almost forgotten the plans for the space station that we had discussed back in 1969. O'Neill's approach, however, was quite different from the one taken by von Braun five years earlier. O'Neill was very much concerned by the negative thinking that characterized the attitude of our best young people in the early 1970s. He told me that he had run across this in teaching his elementary physics courses for the past few years and he was really concerned. He was particularly worried about the then very prevalent preoccupation with the limits that circumscribe what we can do. O'Neill cited the book that was so popular at the time, *The Limits of Growth*, by Donella H. Meadows, Dennis H. Meadows, Jorgen Randers, and William W. Behrens III (Potomac Associates Books, New York, 1972), which was published under the auspices of the Club of Rome. This book was essentially a Malthusian tract that made a relatively sophisticated case for the view that the world's resources were being exhausted and that stern measures of austerity and population control would be required if the human race were to survive. As an added attraction, the book featured a series of calculations based on a computer model that had been developed by Professor Jay Forrester of MIT. There it was, as it were, in black and white for everyone to see; the world was coming to an end—all put on the most modern computer printouts!

I had also become aware of the Club of Rome and of *The Limits of Growth* but in an entirely different connection that I will explain in due course. I shared O'Neill's concerns and I also drew the same political consequences that worried him so much. In our first meeting, O'Neill described eloquently what he thought would happen if the philosophy of limits were actually to dominate the thinking of the world's people. He repeated the argument made in *The Limits of Growth* that much more attention would have to be paid in the future to the distribution of the world's limited resources and then drew the obvious conclusions. If resources were indeed finite, if no more growth was possible, then a political system would have to be invented to eventually divide up the finite pie that was postulated by the limits theorists. It was O'Neill's contention that ultimately the only political system capable of performing this function would have to be authoritarian or fascist in nature. If there was no scope for growth and for expansion, then tyranny was the only way to run society. Someone, in other words, would have to have the authority to say who gets what on a permanent basis and that would lead to fascism and stagnation. It was for this reason that O'Neill felt the then fashionable theory of limits was so dangerous. He went on to say that freedom and a democratic form of government were possible only if people had an instinctive feeling that growth was possible and that there were indeed new horizons to be conquered. I found myself in essential agreement with O'Neill's thinking.

O'Neill then went on to tell me what he had done about the problem as he perceived it. In order to stimulate the imaginations of the students in his freshman physics courses, he gave them the problem of developing a conceptual design for a large space station or space colony as he called it. The students were fascinated, and O'Neill began to take the notion of actually promoting the idea publicly much more seriously. The students also gave their imaginations free rein and, together with O'Neill, they created for themselves the kind of expanding horizon that

O'Neill felt was so important from a political viewpoint. After giving the space colony project to his classes three or four years running, O'Neill decided that the time was ripe for a public statement. In the September 1974 issue of *Physics Today*, there appeared an article on space colonies in which the whole concept was laid out in detail for the first time. The article attracted considerable public attention, and it was for this reason that O'Neill came to see me. What he wanted to do was to organize a thorough engineering "summer study" that would spell out in detail what problems would be faced if such space stations or colonies were actually ever built. Since I shared many of O'Neill's convictions, I agreed to help out. In the summer of 1975, with the help of financial support from NASA headquarters, the first summer study on space colonies was conducted, and O'Neill spent most of the summer at Ames working with the group.

Several important technical ideas were developed during the summer study. One was based on the notion that the construction of large habitats or facilities in earth orbit or somewhere in cislunar space would require much raw material that could be more easily lifted from the moon than from the earth because of the moon's smaller gravitational field. O'Neill had already thought about this problem and, being an accelerator physicist, he developed the idea of an electromagnetic "mass driver" to lift material off the lunar surface. This was essentially a linear homopolar generator that was designed to accelerate specially designed buckets filled with lunar material to lunar escape velocity. The material would then be "caught" out in space somewhere near the space colony by an appropriate device and would be used to manufacture whatever was needed. Enough was known about the nature of lunar material from the Apollo flights that some preliminary guesses could be made about just what could be done with the material. Although the ideas behind manufacture in space did not originate with O'Neill, he greatly expanded the scope of what had been contemplated prior to the summer study in 1975. Finally, there was the question of the actual design of a space colony or a "habitat" that might house as many as 10,000 people. All of this was really quite breathtaking, and the report of the summer study was important because it was the first time that many of the ideas O'Neill was proposing were developed in a quantitative way. The report gave the whole idea of space colonies and the existence of new horizons a sense of reality that it did not possess before. It also confirmed my own conviction that if anything such as this was eventually to be done, then it was necessary to start the process with the construction of a permanently manned, earth-orbiting space station.

Another important event resulting from O'Neill's summer study was the founding of an organization called the "L-5 Society." The purpose of the society was to advocate the colonization of space and the development of the ability to manufacture things in space. It was founded in 1975 by a group of people led by Carolyn and H. Keith Henson. The name of the society was taken from one of the ideas that O'Neill had proposed some years earlier, which was that the most effective place to locate a space colony would be at one of the libration points (or gravity potential minima) of the earth-moon-sun system that had been identified by Lagrange a century and a half earlier. In O'Neill's view, the most convenient place for a space colony would be at the point L-5, hence the name of the society! What was most important about the founding of the L-5 Society was that it marked a turning point in the public attitudes toward the space program. Whereas in the early 1970s the public attitude was either neutral or negative, by the end of the decade there was a strong public feeling that the country should be doing more rather than less in space. Much of the credit of this turnaround in public attitudes was due to organizations such as the L-5 Society that helped to rekindle public interest in space exploration. This public interest would be all-important eight years later when we actually mounted the effort to start the space station program.

Late in 1973 I was drawn into an enterprise that, initially, had nothing to do with the space program but that would ultimately lead to my move to Washington in 1977. My old boss, Edward Teller, had been a friend of Governor Nelson Rockefeller of New York for some years. (Teller held a Rockefeller Foundation grant to support some of his early scientific work in the 1930s.) Sometime in 1973 Rockefeller made the decision to resign the post of governor of New York to which he had just been reelected for an unprecedented fourth term. Rockefeller wanted to devote his time to an effort to rethink the position of the United States in the world from first principles, and he wanted to do this in connection with the bicentennial year that was coming up in 1976. Rockefeller would establish a very broadly based committee of distinguished citizens, and this group would form the focus of the effort to accomplish what Rockefeller had in mind. He called this group the Commission on Critical Choices for Americans. At Teller's suggestion, Rockefeller called me and wanted to know whether I would be interested in becoming the director of the commission's staff. Although I was initially quite intrigued by the prospect, I finally decided not to take the post because I did not feel that my work at Ames was finished. However, I did agree to participate somehow in the work of the commission.

Nelson Rockefeller was a remarkable man. Everything about him was somehow larger than life. When I first met him, he told me that his purpose in establishing the Commission on Critical Choices was to "rewrite the Federalist Papers" because he thought it was a good opportunity to use the bicentennial to "reconvince the American people that the Constitution was a good thing." This was quite a breathtaking concept, and I could not help asking where he would find a John Jay or a James Madison to help him do the job. Rockefeller was not at all fazed by my question—he simply said we will look and we will find them. My own work with the commission was concerned with examining the process of how the development of new technology is managed and applied. I had a number of very interesting discussions with some of the leading members of the commission on these topics and then wound up writing a chapter on technology development for one of the fourteen books that the commission would eventually publish.

While all of this was going on, the tragedy of Watergate was unfolding. As it happened, my family and I spent a few weeks in the summer of 1974 as house guests of the Rockefellers at their Pocantico estate near Tarrytown, New York. I was helping to edit the various papers that commission members had turned in and was also writing my own chapter on technology development. We were there when President Nixon resigned his office on August 9, 1974. The politically astute Rockefeller realized very quickly that the new president, Gerald R. Ford, would ask him to serve as vice president and he began to prepare himself accordingly. Shortly after Nixon's resignation, Rockefeller called a group of us together to help him decide what he should do if and when he assumed the vice presidency. He told us that he wanted to do some things that were of sufficient importance to be worthy of the attention of a political leader of the first rank—which he correctly felt that he was—but that would not cause conflicts with President Ford.

Rockefeller's old friend, the prominent New York attorney, Mr. Oscar Ruebhausen, was in overall charge of this planning process. We decided finally that Rockefeller should concentrate on two things: the rebuilding of the nation's intelligence system to repair the damage done by the Watergate affair and the reestablishment of the Office of the Science Advisor on the White House staff which had been abolished by President Nixon in 1972. Rockefeller took our advice, and as things turned out, he more or less accomplished the objectives that we had outlined for him.

It fell to me to work on the problem of reestablishing the president's science advisory mechanism. Shortly after Rockefeller was sworn in as vice president, he persuaded President Ford to establish a committee called the President's Advisory Group on Science and Technology, which actually consisted of two separate committees, one chaired by Dr. William O. Baker, then the president of Bell Laboratories, and the other one by Dr. Simon Ramo, the vice chairman of TRW. Baker's committee would be concerned with the health of basic scientific research in the country and Ramo's with the development of technology and its application for economic benefit. I was made a member of both committees, and my job was to act as a liaison person between the two groups. I spent a good deal of time in Washington during the Ford administration working with the Baker-Ramo groups and also with various members of the vice president's staff. All of this was a first-class graduate education in politics for me, and I have to confess that I acquired a taste for Washington during those years that turned out to be important later in reaching the decision to leave Ames and move to Washington.

In the meantime, work on the shuttle was progressing. In spite of the fact that I spent a fair amount of time away from home, I was still in charge at Ames and I had work to do. The technical support that we were providing for the Johnson Space Center on shuttle development was continuing and was, fortunately, deemed to be valuable. At the same time, other people at Ames began to look at some of the other problems that might be encountered when the shuttle became operational. A particularly interesting one dealt with the problem of human behavior and capabilities in space. Drs. Joseph C. Sharp, Harold Sandler, and Alan B. Chambers at Ames were the leading figures in this effort. In following their work, I learned that the physical environment aboard the shuttle would be quite benign. For example, the atmosphere in the shuttle cabin would have the same composition as the atmosphere of the earth sea level and in no case would the g-forces in a normal shuttle flight exceed more than twice or three times the force of gravity. In 1975 Sandler and his collaborators conducted a relatively lengthy study to define what might be called the "minimum medical criteria" for flight on the space shuttle. It turned out that these criteria were not very stringent, which meant that almost anyone in reasonably good physical shape could actually fly.

This was an extremely important conclusion for me because it pushed me the rest of the way toward becoming a very strong supporter of the space shuttle program. If it was indeed possible for "ordinary" people to fly, then broad new horizons could be opened by the space shuttle. It would no longer be necessary to become a professional astronaut and to devote many years of one's life to training in order to have the opportunity to fly in space. With relaxed medical criteria and the large margins of safety designed into the shuttle system, it should be possible to fly someone with a month or two preparation and this fact, in turn, opens the possibility of flying scientists, engineers, technicians, journalists, and, yes, even poets who would do their respective things and then share the experience with everyone. In short, the shuttle opened the door for a vast broadening of the human experience in space.

The circumstances I have just described were also apparent to the leadership of NASA. Accordingly, on September 20, 1973, a committee was established by NASA headquarters to look into the criteria that should be employed for selecting people to fly on the shuttle. The group was chaired by Professor Frederick Seitz, who was, at the time, the president of the National Academy of Sciences. The idea was to put together a group of people who could make some judgments about the value of the work that could be done by people who were not trained as astronauts and the risks that would be faced in doing this. The committee consisted of some scientists and engineers as well as some representatives of the manned space flight community. Both Dr. Christopher C. Kraft, who was by then the director of the Johnson Space Center, and I were members of Seitz's committee. As expected, those of us interested in using the shuttle to the utmost pushed for a relaxation in the standards for the selection and training of flight crews and those who had the responsibility to execute the flights (led by Chris

Kraft) resisted the effort to relax standards. Finally, a compromise was struck in which three classes of people who would fly on the shuttle were defined. There were regular astronauts, which included the flight crews, that is, the mission commander and the pilot. These people were the professionals who devoted many years of their lives to training and maintaining proficiency. Safety of flight would be their primary concern. There were mission specialists, also professional space flyers, who would not fly the shuttle, but rather would be trained to perform the operations scheduled for that particular mission. Finally, and it was the creation of this group that represented the real breakthrough in our thinking, there were the payload specialists. Payload specialists would not be professional space flyers; rather, they would fly after a short period of training—a few months at the most—and would fly to tend a particular payload. These people would not be NASA employees and would be selected initially by the owner of the payload, subject, of course, to NASA approval. By creating this category, the door was opened and, as of this writing, the first payload specialist, Mr. Charles Walker, an employee of the McDonnell Douglas Corporation, flew on *Discovery's* first mission launched on August 30, 1984 (STS-41D) to tend the Continuous Flow Electrophoresis experiment.

All of these considerations had turned me into a convinced supporter of the space shuttle by the time I left Ames in 1977. Shortly before leaving Ames, there was a meeting of the local section of the American Vacuum Society at Ames. I was invited to give the keynote address at this meeting, and I took the opportunity to describe the way the shuttle would work and to stress the great importance of selecting the right people to fly on the shuttle. (The speech was printed in the November/December 1977 issue of the *Journal of the American Vacuum Society*.) To illustrate my point in the speech, I assumed that I would myself fly as a payload specialist on the space shuttle and then described the experiments I wanted to do. If I were to give the speech today, I would list a different set of experiments, but my hope that I will eventually be able to fly on the shuttle myself is still alive.

VII

The Air Force and Space

The act establishing the new Office of Science and Technology Policy was passed by Congress in its final version on May 11, 1976. Although the structure of the new office was now rather different than what we had envisaged when Nelson Rockefeller asked us to look at the problem shortly before he became vice president, this was still a significant step forward. There would be, once again, a science advisor to the president so that high level technical advice would be available in the White House. There would be no advisory committee such as the old President's Scientific Advisory Committee (PSAC), which was the centerpiece of the old scientific advisory structure abolished by President Nixon in 1972. On this point, the Congress agreed with our suggestion that the PSAC had not functioned as a confidential advisory body but, rather, had, on a number of occasions, been used by various committee members as a public platform to achieve their own ends. Thus, the Baker-Ramo advisory group went out of existence once the new Office of Science and Technology Policy was established.

On July 22, 1976, President Ford appointed Dr. H. Guyford Stever, then serving as director of the National Science Foundation, to become the first director of the Office of Science and Technology Policy and the new science advisor to the president. Guy Stever is an extremely distinguished scientist, engineer, and public servant who had actually been informally acting as the science advisor in his post as director of the NSF—an arrangement that President Nixon had initiated after the first science advisory structure was dismantled. However, the new statutory Office of Science and Technology Policy required a full-time director and, therefore, Stever had to resign his post as director of the NSF.

Sometime in July 1976 I had a telephone conversation with Vice President Rockefeller during which he asked me whether I would be interested in succeeding Stever as the director of the NSF. He told me that President Ford, who I had met a few times during my service on the Baker-Ramo group, was very interested in having me take the post. Rockefeller's proposition was interesting to me, and I told him that I would come to Washington and talk with the relevant people about it. I went to see Guy Stever, who was still in his office at the NSF, to ask him some questions about the job. Our meeting was a little embarrassing to both of us since no one had apparently informed Stever that I was being considered as his successor. Stever was very gracious about this business and, among other things, suggested that I talk with Dr. Norman Hackerman, the former president of Rice University, who was then serving as the chairman of the National Science Board, the body that oversees the operation of the NSF. Both Stever and Hackerman advised me not to accept the appointment. They argued that I would risk becoming a political problem by stepping into the director's job shortly before a presidential election. If President Ford were reelected, well and good; but if he were not, I would be in a really awkward position. Since the Senate could not act on my appointment before the election, a new president might not want me and might therefore not resubmit my nomination to the Senate. This would not only be embarrassing to me but would also politicize the NSF director's job in a way that none of us really wanted. I quickly agreed with Stever and Hackerman and told Rockefeller that I would be pleased to serve as NSF director if President Ford were reelected, but that I could not accept the nomination before the election. Rockefeller agreed that this was the right decision under the circumstances. What I began to realize during this episode was that, somewhat to my own surprise, I had contracted a serious case of "Potomac Fever" and that I very much wanted to spend some time serving in Washington in a capacity where I could contribute to the national technology development effort.

When former Georgia governor Jimmy Carter won the 1976 presidential election, all of the plans I have just described became null and void, and I resigned myself to staying in California. What changed things once again was the selection of Dr. Harold Brown, then the president of the California Institute of Technology, as secretary of defense by the president-elect. I had worked closely with Brown on the Argus experiment twenty years earlier (see chapter 3) but did not have too much contact with him afterward. Although in 1960 it was Brown who, while he was serving as the director of the Livermore Laboratory, hired me to head P-Division (see chapter 4), he had left the laboratory shortly thereafter to succeed Herbert York as the director of defense research and engineering in Washington (1961). After Brown's nomination as secretary of defense, I wrote him a letter congratulating him and offering to help him as best I could. In due course, I received a telephone call from Dr. Eugene G. Fubini asking whether I would be interested in serving in the new administration as under secretary of the Air Force. Fubini was an old friend with whom I had served on a number of advisory committees. (Fubini was a member of Ramo's committee of President Ford's Advisory Group on Science and Technology.) He was, at the time, helping Brown to put together his senior staff in the Pentagon, and he strongly urged me to accept the post, which I did on April 1, 1977.

Aside from being the second ranking civilian in the U.S. Air Force hierarchy, the under secretary of the Air Force has overall responsibility for overseeing the military space program. It is worth quoting the job description that appears in the 1976 issue of the *U.S. Government Manual*: "The Department of the Air Force is administered by the Secretary of the Air Force who is responsible for and has the authority to conduct all affairs of the Department. His principal assistant is the Under Secretary who acts with full authority of the Secretary on all affairs of the Department. The Under Secretary is specifically responsible for overall direction, guidance, and supervision of the space programs and the space activities of the Air Force."

I would therefore continue to be active in the space business and, because of the eight and a half years I had just spent at Ames, I might be of use in helping the Air Force to learn how to use the space shuttle and the other new things that were being developed by NASA.

In order to understand the situation I faced when I began my service in the Air Force, it might be valuable to outline very briefly the history of Air Force activities in space. The military had, of course, been active in space operations from the very beginning. *Explorer I*, the first U.S. earth orbiting satellite, was launched using a rocket adapted from the design of the U.S. Army's Jupiter Intermediate Range Ballistic Missile (IRBM). The Jupiter rocket was put together by Wernher von

Braun and his team during the 1950s working at the U.S. Army Ballistic Missile Agency (AMBA) located at the Redstone Arsenal at Huntsville, Alabama. (Later, in 1961, part of the arsenal became NASA's George C. Marshall Space Flight Center.) Furthermore, John Glenn made his first orbital flight in 1962 using a modified Air Force Atlas Intercontinental Ballistic Missile (ICBM) and a modified Air Force Titan ICBM was used for the Gemini program. The military was, therefore, involved in the first instance because they had developed the launch vehicles. (It is an interesting fact that, with the exception of the Saturn and the shuttle, all other U.S. space launch vehicles—the Titan, the Atlas, the Thor-Delta, and the Jupiter—were originally developed by the military.) However, they also were interested in satellites and the U.S. Army launched the first military communications satellite, *Score*, on December 18, 1958. *Score* had an active transducer to receive and amplify the return signals. *Score* was followed by *Echo* (1960), *Relay* (1962), and *Syncom* (1963)—all of which were developed by NASA. The Air Force launched the first *Discoverer* satellite on February 28, 1959, and the Navy had an operational navigation satellite system, *Transit*, by 1964.

The leader of the early Air Force space effort was the remarkable General Bernard M. Schriever. In the 1950s Schriever had been responsible for organizing the Air Force ICBM program so that he was thoroughly familiar with what could be done. During the 1960s Schriever was involved in the planning for two large Air Force space programs that were then on the drawing boards. One of these was called Dyna-Soar, which was a small reusable "space aeroplane" designed to be launched using a Titan rocket that could be used to bring people back and forth from near-earth orbit. The other was the "Manned Orbiting Laboratory," or MOL, which was a small three-person space station. The MOL was also designed to be launched by an advanced Titan launch vehicle (the so-called seven-segment Titan 3C), and it would eventually be tended by something like the Dyna-Soar. Substantial sums of money (that is, several billion dollars) were spent on both of these projects, and Schriever, along with some colleagues, was beginning to develop the military doctrines that would eventually govern space operations related to the national security.

Unfortunately, the military—and particularly the Air Force—suffered two serious setbacks in the 1960s that caused Schriever's plans to lie dormant for more than a decade. The success of NASA's Mercury and Gemini programs in the early and mid-1960s precipitated a review of military space operations. Since Dyna-Soar was technically very ambitious, the general feeling was that the nation's manned space flight effort was adequately covered by the NASA program and that what was learned by NASA could eventually be adapted by the military. Accordingly, Defense Secretary McNamara ordered the Dyna-Soar program to be canceled on December 10, 1963. However, the plans for the MOL program survived.

In the late 1960s, just as the time to turn the plans for MOL into reality approached, the post-Apollo planning effort in NASA was getting under way. Also, the same argument developed within the military that was going on in the civilian space program, that is, the debate over the value of sending people into space. There were (and are) people in the military and the national security related space business who questioned the value of sending people into space and who maintained that all of the important military requirements could be better fulfilled using automated spacecraft. They maintained that, in addition to being less expensive, unmanned spacecraft would be less vulnerable to interdiction or destruction. The reasoning went that anything with people on it would be more valuable a priori, and we would therefore be reluctant to launch or operate it in crisis situations. At the same time, the post-Apollo planning process was yielding the plans that would eventually lead to the shuttle and to the space station. Apparently there were some people in the administration who felt that NASA's plans would duplicate what the military had in mind and that one or the other would have to be canceled. The upshot of these considerations was that the MOL program was cancelled on June 10, 1969. Thus, the people in the national security establishment who advocated the development and use solely of unmanned spacecraft for national security purposes gained the upper hand. They made a virtue of necessity and, as we shall see, have a dominant voice in military space matters to this day.

There were, of course, numerous people in the military who argued vigorously against the cancellation of the manned space flight effort in the military. Among these were the people who had actually been selected to fly on the MOL as astronauts including two young Air Force majors, James Abrahamson and Robert Herres, who would subsequently, and in different ways, make very important contributions to the nation's space program—of which more later.

At the time, there were many people both in the national security establishment and elsewhere who felt that, in spite of the separation between the civil and military space programs written into the 1958 Space Act, the Air Force should not be completely left out of manned space flight. To accomplish this objective, it was decided some time in 1971 that the Air Force would become a partner in the space shuttle program and that Air Force requirements would be taken into account. There were people in NASA who were not happy with the proposed arrangement. As the discussions progressed, it was apparent that the Air Force would require a large "cross range" for the space shuttle (that is, the ability of the shuttle to land at a point some distance from the point at which the orbital plane cuts the earth's surface), perhaps as much as 1,000 nautical miles and that the payload capacity of the shuttle would have to be large enough to accommodate the largest national security related payloads then on the drawing boards. In order to deal with these requirements, the straight wing configuration for the Orbiter that was favored by many in NASA at the time was abandoned and the current delta wing configuration was adopted. This change provided the cross range capability that the Air Force required. (The cross range requirement arose from the desire of the Air Force to be able to launch the shuttle into a polar orbit, deploy a payload, and return to the original landing site in the same orbit. A simple estimate will show that a cross range of about 1,000 nautical miles is required to execute this maneuver.) To meet the payload requirement, the shuttle Orbiter was made somewhat larger than the NASA planners originally had in mind.

In 1972 an arrangement was made that clearly defined the NASA and the Air Force roles in the shuttle program. NASA would be responsible for the Orbiter, the solid rocket boosters, the large external tanks, and the launch site on the east coast at the Kennedy Space Center. The Air Force would be responsible for developing the upper stages used by the shuttle and the west coast launch site then planned for the Vandenberg Air Force Base. It was furthermore agreed that the funding for the Air Force portion of the program would be carried in the Air Force budget. The cost estimates at the time were that the NASA portion of the program would cost around $6 billion and the Air Force portion would cost about $2 billion. These estimates, as it turned out, were quite optimistic.

In 1972 when the NASA-Air Force agreement was made, the people in the Air Force who were actually conducting space operations were not much affected. The development of the shuttle was a long-term effort, and the time to accommodate Air Force operations to the shuttle was still some years in the future. The leadership of the Air Force, that is, the secretary and the senior military people, did support the agreement. Probably the strongest support in the defense establishment for working with NASA came from the office of the secretary of defense through the director of defense research and engineering. The original agreement was made when Dr. John S. Foster, Jr. (another former director of the Livermore Laboratory) held that position, and Foster was then and is now a strong supporter of the space shuttle program. His successor, Dr. Malcolm Currie continued this support. The program coordination was accomplished through the Aeronautics and Astronautics Coordinating Board, which was (and is) cochaired by the director of defense research and engineering (now the under secretary of defense for research and engineering) and the deputy administrator of NASA.

I arrived in Washington on July 8, 1977, to assume my full-time duties as under secretary of the Air Force. My boss was Mr. John C. Stetson of Chicago, Illinois, the new secretary of the Air Force. Stetson is an extremely shrewd judge of events who even then foresaw that there would be serious trouble of a military nature in the Persian Gulf and who persuaded the Pentagon's military analysts to put a section on the Persian Gulf in the strategic planning documents. Stetson was also a very strong supporter of the Air Force space program and provided critical help to me on a number of occasions in that area. My other colleagues among the civilian Air Force secretariat were Mrs. Antonia Handler Chayes, (assistant secretary for manpower and reserve affairs), Dr. John J. Martin (assistant secretary for research, development and logistics) and Mr. Jack Hewitt (assistant secretary for financial management). I was finally confirmed by the Senate on July 22, 1977, and sworn in shortly thereafter.

On August 12, 1977, the first Approach and Landing Test (ALT) of the space shuttle was conducted by astronauts Fred Haize and Gordon Fullerton at Edwards Air Force Base. In this test series, the space shuttle *Enterprise*, which is essentially a structural mockup of the shuttle Orbiter without working propulsion or thermal protection systems, is taken to altitude on the Boeing 747 shuttle carrier aircraft. It is then released and flown to a landing. (The *Enterprise* does have a working flight control system.) NASA took the opportunity to make an occasion out of this event, and the festivities during the landing gave me a good opportunity to talk about the shuttle with some of my new friends in the Air Force who also attended the landing. Nothing of substance was discussed, but it was fortunate that the first Approach and Landing Test Mission occurred so soon after my start in office because it did give me the opportunity to draw attention to the shuttle and what could be done with it. The flight itself was a spectacular success. It was really impressive to see the large white Orbiter separate from the shuttle carrier aircraft and then come to a smooth landing on the dry lake bed. The reaction from the crowd that was present on that day foreshadowed the great public excitement that would be generated by the first real space flights of the Orbiter four years later.

The new leadership in the Pentagon continued to support the shuttle program. Both Harold Brown and Dr. William J. Perry, who became the new director of defense research and engineering, were very knowledgeable about the state of the shuttle program and about the major issues before us in the military space effort. The appointment of the latter was particularly fortunate for me. Perry had been a friend of some years standing and, during the years I was at Ames, we both liked to fly on the night airplane from San Francisco to Washington (the "red-eye special") when we had business there. We often had drinks and long talks on these flights. Perry's company, Electromagnetic Systems Laboratories, which he founded and headed for some years, was located in an industrial park right next to the Ames Research Center, so we were neighbors as well. Perry also had asked me to serve on the technical advisory committee of the Defense Intelligence Agency (DIA), which he chaired at the time so we had many opportunities to exchange views and to get to know each other's thinking. Both Brown and Perry were to play very important roles in keeping the space shuttle program on track during the Carter years.

While the senior civilian leadership continued to support the space shuttle program, the people in the Air Force who actually had the responsibility for conducting the national security related space program had a much more guarded attitude. One of the conditions that

President Nixon's Office of Management and Budget (OMB) had made in exchange for approving the shuttle program in 1972 was that the shuttle would eventually replace all other launch vehicles. This was an important part of the "economic" argument made to justify the space shuttle, and it was controversial then and is still a source of controversy to which I will return in later chapters. In any event, the philosophy that the Air Force operational people had evolved toward the shuttle consisted of two major points:

1. The shuttle would be regarded primarily as a "truck" and that the interfaces between the shuttle and the payload would be minimized.

2. The capability to launch national security related payloads on conventional expendable launch vehicles would be retained until such time that the shuttle was proved to be completely reliable.

On the face of it, this policy was quite prudent and reasonable. The great political and military importance of the national security related space program was such that no chances should be taken with respect to the ability to get these payloads into earth orbit. (The people who launch commercial communications satellites adopted essentially the same view and also retained the capability to launch their payloads on conventional launch vehicles.) There was, however, a technical price that had to be paid for the maintenance of this policy. The spacecraft had to be designed in such a way that they were "dual compatible," that is, they would be flown both on the space shuttle and on the expendable launch vehicles. Since the space shuttle was much more capable as a launch vehicle than the conventional rockets, the spacecraft was compromised in terms of capability. Also, and this point would become more important as time went on, by minimizing the interfaces with the space shuttle, the possibility of using crew members to tend the spacecraft and to be part of the operation was abandoned.

During my service in the Pentagon from 1977 to 1981, I tried to modify the policies of the Air Force toward the space shuttle. One thing I tried to do was to urge people to design their spacecraft in such a way that full advantage would be taken of the capability of the space shuttle. I was partially successful in doing this, and certain spacecraft were designed to take full advantage of the payload capacity of the shuttle and of the volume of the payload bay. (It is interesting that seven years earlier the design of the shuttle was, of course, developed in such a way that just these things could be done.) In addition, I also succeeded in getting some of the people in the Air Force to think about the possibility of building their spacecraft in such a way that they could be retrieved and then refurbished and used again. There was even the possibility of repairing, replenishing, and maintaining spacecraft on orbit by using the ability of the shuttle crews to go out and perform extravehicular activities (EVA's). These things are now slowly being done, and I have to admit that I underestimated the difficulties at the time that I made these suggestions. It would take more than six years before the retrieval and repair of a satellite on orbit would be demonstrated by the crew of the STS-41C mission in April 1984.

On balance, I believe that the conservative attitude of the Air Force toward the space shuttle at the time was probably justified. We were to encounter delays and problems in the space shuttle program that would indeed call for a cautious approach. Perhaps the most articulate exponent of the Air Force position at that time was Mr. Jimmie D. Hill, who was then a member of the under secretary's staff and who would later become the deputy under secretary of the Air Force for space systems. Hill had an encyclopedic knowledge of Air Force space systems as well as a first-class intelligence that he applied to the problems at hand. By taking positions that were generally opposed to mine, we usually arrived at workable compromises that could be implemented. Hill became a trusted and valued associate during my years in the Pentagon. Another individual who had considerable influence on my thinking during my first year in Washington was my first military assistant, Colonel (now Lieutenant General) Harry A. Goodall. Goodall is a brilliant combat officer who also possesses a first-class analytical mind. Although he had never before had any dealings with space operations, he mastered the essentials in an astonishingly short period. Whereas Hill would represent a generally conservative approach, Goodall, who was not tied to the existing thinking about space operations, was usually willing to support the more "radical" things that I was proposing. Hill and Goodall together provided for me the best possible support that I could expect during my first year of service in the Pentagon.

During the festivities that surrounded the first Approach and Landing Test of *Enterprise*, in August 1977, I was fortunate to be able to renew my friendship with Tom Stafford. Stafford had left the astronaut corps after the completion of the Apollo-Soyuz program to return to active duty in the Air Force. In August 1977 he was serving as commander of the Air Force Flight Test Center at Edwards Air Force Base and had by now been promoted to the rank of major general. Stafford is a person of great courage and intelligence who was one of the most respected of the astronauts. He had participated in the Gemini (*Gemini 6* with Wally Schirra and *Gemini 9* with Gene Cernan) and in the Apollo (*Apollo 10* with Gene Cernan and John Young) programs and had served as the commander of the Apollo-Soyuz mission. He was one of the most experienced astronauts, having logged more time in space than anyone except John Young. I had a long talk with Stafford, and we quickly agreed that the attitude of many of the Air Force people toward manned space flight and the use of the space shuttle was not very productive. During our meeting, Stafford agreed that he would be willing to come to Washington to serve in some appropriate capacity on the Air Staff. In due course, this was indeed accomplished. In March 1978 Lieutenant General Alton D. Slay, who was then serving as the deputy chief of staff for research and development in the Pentagon, was promoted to the rank of general and placed in command of the Air Force Systems Command. I went to the then chief of staff of the Air Force, General David C. Jones, and suggested that Stafford be selected to replace Slay. My advice was accepted, and Stafford became the new deputy chief of staff of the Air Force for research and development along with a promotion to the rank of lieutenant general on March 30, 1978.

Stafford's presence in Washington was important to me because I now had at least one high-ranking ally in the uniformed Air Force who also felt that manned space operations would eventually be important to the Air Force. Shortly after he arrived in Washington, Stafford and I decided that it might be valuable to conduct a formal study of what could be done for the Air Force and the military in general with manned space operations. Harold Brown agreed to sponsor the study so that the other military services would also be represented. Stafford and I then organized the study groups with myself heading the overall "Steering Group" and Stafford heading the "Working Group" for the conduct of the work involved in producing the study. We decided to call the study "The Utility of Military Man in Space," and we divided the work into various time periods, near-term, medium-term, and long-term, during which we thought that various manned military operations might become important. A large number of people were involved in the study, and if nothing else the conduct of the study had the effect of sensitizing many people to the possibilities inherent in manned space flight for military purposes. The study documents are classified, and it is not really necessary for the purpose of this book to discuss some of the details of the considerations that were included in the study. We did consider in some detail what might be accomplished with a space station and concluded that in the longer term a space station could be valuable for military purposes.

The "Utility of Military Man in Space" took over a year to complete. The final version of the study was submitted to the secretary of defense on October 24, 1978. As it turned out, the study had little impact because it was—as so many such efforts are—premature. When the study was submitted, the shuttle program was in some difficulty, and most of us devoted considerable time and effort to keep the shuttle program from being truncated or, at worst, canceled. Nevertheless, the effort that was put into the study did pay off, in a more limited way, in that it stimulated many people to think about the problems of manned space flight and the possible military applications. At some point in the future, I have no doubt that at least some of the things that were proposed in the study will be implemented.

Late in 1977 I made my first visit to the headquarters of the North American Air Defense Command (NORAD/ADCOM) in Colorado Springs. At the time, General Daniel E. (Chappie) James, Jr., was serving as the commander in chief. By the time I made my visit, James had already asked to be relieved of his post as commander in chief because of ill health (unfortunately, this very distinguished Air Force officer died at the early age of fifty-eight shortly after he left the command). General James was replaced by General James E. Hill on December 6, 1977. I made another trip to Colorado Springs to visit General Hill at his headquarters on April 21, 1978. At the time, the chief of staff, General Jones, was proposing to make a comprehensive reorganization of the North American Air Defense Command. The interceptor squadrons (mostly F-102 and F-106 aircraft) that were at that time dedicated to air defense and assigned to the Air Defense Command would be taken away from the command and assigned to the Tactical Air Command (TAC). The warning systems, that is, the radars and the satellites designed to provide the signals that would provide warning that a bomber or an ICBM attack had been launched, would be turned over to the Strategic Air Command (SAC). The Air Defense Command would retain only the "operational" control over these forces but would not have any direct management control. In my view, this was not a satisfactory arrangement, and I argued strongly against the proposal. I discussed this problem with General Hill during our meeting. We agreed that a new mission had to be provided for the command that would revitalize the organization and permit the new commander in chief to bring in some good new young officers to the command. The best new mission, we believed, would be to organize an Air Force Space Command as part of the Air Defense Command and to turn over the management and the operational control of all Air Force satellite systems to the Space Command. Since the two primary functions of the satellites were attack warning and communications, General Hill and I both felt that it would be natural to tie the Space Command to the Air (and missile) Defense mission. This was something that had already been advocated by General Bernard A. Schriever some years earlier, and General Hill and I decided at that meeting to revive the idea and to push it as best we could. It would be more than five years before the Space Command as we then conceived it would become a reality.

Early in the Carter administration an attempt was made to cut back the space shuttle program, which was at least partially successful. The original proposal came from the analytical staff of the Office of Management and Budget (OMB), but it was later also picked up by some people in the Congress who asked the General Accounting Office (GAO) to look at the possibility of cutting back the shuttle program. At that time, the plan was to build five shuttle Orbiters and two launch sites, one on the east coast (Kennedy Space Center) and another on the west coast (Vandenberg Air Force Base). On November 29, 1977, I was invited to attend a meeting in the office of Mr. James McIntyre at which the future of the space shuttle program would be discussed. McIntyre was, at that time, serving as acting director of OMB after the resignation of Mr. Bert Lance, President Carter's first budget director. (McIntyre would subsequently be appointed to succeed Lance and would serve for the remaining years of the Carter presidency.) It was a high-level meeting—Harold Brown, Charles Duncan (the deputy secretary of defense), Admiral Stansfield Turner (the director of central intelligence), and William Perry were all present. McIntyre's staff presented us with three options for the future of the space shuttle program. One was to stick with the current plan (five Orbiters and two launch sites). The second was to cut back the shuttle program to an experimental one in which only three Orbiters would be completed and only one launch site—the one on the east coast—would be activated. (Under this option, the plan to ultimately use the shuttle as the exclusive launch vehicle for all payloads would also be abandoned.) Finally, there was a third option that would be a compromise calling for the construction of four Orbiters and would leave the question of launch sites open. The OMB spokesman made a strong case for the three Orbiter, one launch site option on the grounds that this would help the administration's near-term budget problems. There was a general discussion during which I argued that we should stay with the original plan and during which it became apparent that both Perry and Brown would not go along with the option favored by OMB. No decisions were reached at this meeting.

On December 16, 1977, there was another meeting with McIntyre, Brown, and Turner to discuss the national security related space program, as well as the future of the space shuttle. This meeting was called to make some final decisions about the president's Fiscal Year 1979 budget. It was at this meeting that Harold Brown took a very strong position that the OMB proposal (three Orbiters and one launch site) was unacceptable from the viewpoint of national security. He made the case that at least two launch sites (one on the east coast and the other on the west coast) would be required and that at least four Orbiters would be necessary to meet the requirements of national security. This last argument was based on the fact that the first two Orbiters to be built (OV-102, *Columbia*, and OV-099, *Challenger*) would be somewhat heavier than the following vehicles and would therefore not be capable of carrying the very heaviest national security related payloads. It was therefore necessary to have at least two Orbiters capable of carrying the very heaviest payloads in order to have a backup in case one of these vehicles was lost. This argument carried the day and the decision was reached to build four Orbiters (OV-103, *Discovery*, and OV-104, *Atlantis*, in addition to the first two) and to continue with the construction of the west coast launch site. (The west coast launch site was deemed necessary in order to conduct polar orbiting flights required for national security related missions.) Admiral Turner supported Brown's position. This was the first, but not the last, time that Brown would help rescue the shuttle program when it was under attack. The essential outcome of the debate over the Fiscal Year 1979 budget was therefore that the original shuttle program would be modified in that only four, rather than five, Orbiters would be built, but that the plan to construct two launch sites would remain intact.

The argument over the structure of the shuttle program would be repeated when the President's Fiscal Year 1979 budget was submitted to the Congress. Some analysts at the General Accounting Office had taken the original OMB plan and studied it to see whether some of the arguments related to national security requirements could be answered in another way. They had hired a technical consultant and had developed a plan for a way to conduct polar orbit missions from the east coast launch site. They conceded the requirement for four Orbiters but argued that it would be possible to fly "dog-leg" trajectories from Cape Canaveral that would make polar orbit flights possible. These "dog-legs" involved turning north shortly after launch and flying the first part of the polar orbit trajectory over the continental United States. The GAO analysts took the position that the risk in doing this would be acceptable in view of the relatively small number of polar orbit missions then being planned. On February 13, 1978, I met with Richard Gutman, who was the principal GAO official in charge of the shuttle analysis, to make the case that the "dog-leg" suggestion would not be acceptable and that we could not risk being forbidden to conduct such operations because of public objections to the proposed overflights of the United States. This argument eventually carried the day, and the funds to continue the construction of the Vandenberg launch site in the Fiscal Year 1979 budget were approved by the Congress. Nevertheless, in spite of this success, the construction of the Vandenberg launch site would continue to be a contentious issue in the coming years.

In December 1977 I was invited to deliver the principal speech at the annual Goddard Memorial Dinner sponsored by the National Space Club scheduled for March 10, 1978. I was very pleased with this invitation and took the opportunity to make a more or less visionary speech on what the future held for the space program. I pointed out that we were generally too pessimistic in making predictions about what was possible in the future. Accordingly, I included three speculations about what would happen by the year 2000. One was that many hundreds—maybe even thousands of people would fly in space, including some that were then sitting in the audience, before A.D. 2000. A second was that we would establish a space station on which many people would be living permanently, and the third was that we would find evidence of extraterrestrial intelligent life. It will be interesting to see if any of these speculations actually turn into real events.

VIII

Space Policy, Arms Control, and Organizational Problems

The National Aeronautics and Space Act of 1958 is a truly remarkable document. It calls for the establishment of a space program managed by civilians, dedicated to exploration and the development of new technology, and carried out under a broad charter to provide information for the public on what is being done. The same act assigns to the Department of Defense the management of those space programs related to the national security. Finally, the act requires that the civilian (i.e., NASA) and the military space programs share technology to the advantage of both parties. The creation of two separate space programs by this act has turned out to be very successful, and the arrangement has served the nation well. We have been able to operate a highly publicized and very successful space program through NASA, which has had important symbolic as well as substantive value because it is completely open to the public. Simultaneously, we have been able to conduct an equally successful classified program under the auspices of the Air Force designed to fulfill very important national security requirements. I have had the good fortune to work on both sides of the U.S. space program and, based on the experiences I have had, I believe that the arrangement specified by the 1958 Space Act is an excellent one.

Since the Space Act establishes two separately managed space programs, one civilian and one military, some mechanism must exist to adjudicate disputes that will inevitably arise. Since both the secretary of defense and the administrator of NASA report to the president, the only place where such adjudication can occur is in the White House itself. During my years in Washington, a mechanism has evolved in the White House among the staff people in the Office of Management and Budget, the National Security Council, and the Office of Science and Technology Policy for coordinating and overseeing the civil and military space programs. I cannot say that this mechanism is perfect, but it did serve the purpose for which it was intended. After the staff people finish their analyses, there are usually one or two very senior people who have taken an interest in the space program and who also have access to the president who will then make the final judgments. In the Carter administration, the two most important of these were Mr. David Aaron, the deputy national security advisor and Mr. W. Bowman Cutter, the deputy director of OMB. This coordinating process is, of course, not confined to the space program since there are many other issues requiring some adjudication in the White House. Therefore, it is necessary for each administration to quickly develop methods for accomplishing this objective.

In the Carter administration, the method was to define various problems requiring coordination by a Policy Review Memorandum (PRM —pronounced "prim," of course, according to the local jargon in Washington!), which was prepared by the White House staff. A Policy Review Committee (PRC) was then established to deal with the issue and to recommend an appropriate course of action to the president. The result of the process was then usually summarized in another document and this was called the Presidential Decision (PD). (It is amusing that, four years later, exactly the same kind of system was evolved by President Reagan's staff—except that the acronyms that describe the committees and the documents were different! I will explain these, in turn, at the appropriate place.)

On August 4, 1977, less than a month after I arrived in Washington, I went to a meeting of a group that was concerned with the development of the Carter administration's space policy. A Policy Review Memorandum (PRM-23) had been issued calling for a space policy and the purpose of the meeting was to establish the ways and means. It was a high-level group with Harold Brown in the chair and three other cabinet officers present (Mr. Cecil Andrus, the secretary of the interior, Mr. Robert Berglund, the secretary of agriculture, and national security advisor, Dr. Zbigniew Brzezinski). The other interested parties (commerce, the CIA, and others) were represented by the appropriate deputies. Although it was a high-level meeting, the decision taken was not a controversial one. It was simply to establish a Policy Review Committee (Space) with the appropriate representation to develop a space policy for the administration. The PRC (Space) would be chaired by President Carter's science advisor, Dr. Frank Press. This decision was an important one because Press was close to the president and would turn out to be very influential in the administration. (It was largely due to the efforts of Frank Press that the decline of the support of basic research by the federal government that had been going on since 1967 would be halted and turned around in the Carter years.) As things turned out, I would not become a member of the PRC (Space), and this was a real disappointment because I thought that I would therefore not have a direct influence on the shape of the administration's space policy.

At about the same time, another Policy Review Committee was established to coordinate the activities of the nation's intelligence activities. This was the PRC (Intelligence) or PRC (I) and its function was to oversee and coordinate the military and civilian portions of the intelligence programs. The director of central intelligence, Admiral Turner, chaired this group. The PRC (I) was important for the space program because some of the national security related space programs for which I was responsible had to do with surveillance functions that fell under the jurisdiction of the PRC (I). The PRC (I) went into full operation in the fall of 1977 and, once again, I would not be able to sit on the committee either, even though the decisions before the PRC (I) would have very important consequences for the budgetary plans I would have to make.

While I was disappointed that I would not be a member of either of these policy-making committees, I was able to establish relationships with people who were close to the operation of both groups in such a way that I was able, more or less, to keep up with what was going on. The chief staff member of the National Security Council for Space and Intelligence matters was Colonel (now Major General) Robert A. Rosenberg, USAF. Rosenberg was extremely knowledgeable and experienced both in space and in intelligence matters, and he quickly became (and still is) a close friend. He was responsible for the organization of the PRC (I) meetings, and he was therefore completely familiar with what was happening in the deliberations and was able to keep me informed. Rosenberg was extremely careful to make sure that I would only be informed of those things for which I had a legitimate responsibility and never talked about things not of concern to me. Because of his discretion and intelligence, Rosenberg commanded everyone's respect and turned out to be extremely effective.

In the case of the PRC (Space), my old NASA friends were very helpful. Dr. Robert Frosch had been appointed as the new NASA administrator at the start of the Carter administration. I was slightly acquainted with him during his term of service as assistant secretary of the Navy during the Johnson administration. Dr. Alan M. Lovelace stayed on as deputy administrator in the new administration, having been appointed in 1975 to succeed George Low when the latter left NASA to become president of the Rensselaer Polytechnic Institute. I was at least partially responsible for persuading Lovelace to join NASA in early 1974 when he took the post of associate administrator for aeronautics and space technology. In addition to good relations with the NASA leadership, I was fortunate to be able to establish a good personal rapport with Frank Press so that I really had all of the necessary leverage to influence things in which I had an interest—at least to some degree.

My major concern—both with respect to the civil and the military community interested in space operations—was to make sure that the commitment to pushing through the shuttle with four Orbiters and two launch sites would be maintained. This was not always easy to do because there was considerable opposition, both in the military and elsewhere, to a policy that called eventually for exclusive dependence on the shuttle for all launches. I actually was in agreement with the reservations that were expressed on this point at the time. On the other hand, the shuttle program was not far enough along so that it was completely safe, and I therefore swallowed my reservations and became a very strong advocate for a complete and comprehensive commitment to the completion of the space shuttle. Once the shuttle became operational, there would be plenty of time to see whether other launch vehicles than the space shuttle should be maintained or developed.

Early in August of 1978 I completed a lengthy report for Harold Brown on the status of the space shuttle in which I strongly recommended that the Defense Department continue to support the shuttle program. I was convinced (and still am) that the enormous and flexible capability that the shuttle represents would ultimately be very important for the national security. In this report, I also discussed the status of the west coast launch site construction and began to build the case that the Defense Department—with the Air Force as the executive agent—should begin to develop the ability to perform mission control functions. Brown accepted these recommendations and the commitment to use the shuttle remained in force during the Carter years.

Since there were two policy committees dealing with space, the PRC (I) and the PRC (Space), it is not surprising that two policy documents were eventually issued on space by the Carter administration. The first of these was developed by the PRC (I) and was drafted initially by Colonel Rosenberg. Mr. Benjamin Huberman, who served both on the staff of the National Security Council and the Office of Science and Technology Policy also made extremely important contributions. The document was issued in its final form on May 11, 1978, and was labeled PD-37. It is a highly classified paper that has as its major purpose a clear statement on the division of responsibility between the secretary of defense and the director of central intelligence for the development and use of data gathered from space. The document also contained a strong endorsement for the shuttle program. Both from the viewpoint of my Air Force responsibilities and my support of the shuttle program, PD-37 was entirely satisfactory.

The document on space policy being prepared by the PRC (Space) was labeled PD-42, and it was issued on October 10, 1978. PD-42 was more broadly based than PD-37 and was mostly an unclassified document that touched upon all of the aspects of the nation's space activities, scientific, commercial, and those related to national security. A copy of the unclassified portions of PD-42 is reproduced in appendix 2.

Although I was not formally a member of the PRC (Space), I was consulted frequently on the content of PD-42 both by the people in the Defense Department responsible for policy development and by other members of the PRC (Space). In the Defense Department, I worked mostly with retired Admiral Daniel J. Murphy who was the deputy under secretary of defense for policy with the overall responsibility for space policy matters in the Office of the Secretary of Defense. Murphy is a shrewd, able, and experienced official with whom I had many discussions on space policy during the years I served in the Pentagon. Although Murphy was skeptical about the value of the space shuttle initially, he eventually became a strong supporter. PD-42 does contain a statement of support for the shuttle program, and this was, of course, one of the things that I wanted to see in the document.

On September 15, 1978, Frank Press sent me a draft of PD-42 for review. The document called for a broadly balanced space program with the proper stress on the importance of the national security related space efforts. There was also an emphasis on space applications, which I knew Bob Frosch believed to be very important. There was one statement in PD-42 with which I strongly disagreed, and it is quoted directly as follows: "It is neither feasible nor necessary at this time to commit the US to a high challenge, highly visible space engineering initiative comparable to Apollo." I did not believe that it was wise to include such a statement. I was worried that an explicit policy pronouncement of this kind would eventually make it more difficult to initiate a space station program, which was something I thought we would want to do in the coming years. I remember that a few days after reviewing PD-42 my wife and I took Press and his wife sailing on our boat. During our afternoon's cruise on the Chesapeake Bay, I had the opportunity to make my case. As a matter of principle, I told Press, I thought it unwise to include negative statements in a policy statement of this kind. If you are *not* going to do something, then you just don't do it, but you don't have to say so. Press, in turn, made his case by saying that he wanted the statement in the policy because he was concerned about the continuing commitment to the shuttle. He said that President Carter was a frugal man and that he (Press) was afraid that if the shuttle did indeed get into technical difficulties that the president might be tempted to cancel the program. The negative statement about large new initiatives was necessary, Press felt, so as not to divert attention from the completion of the shuttle program. While I could see Press's point, I could not really quite agree with it. Two years later, I would be put in a position where I could not avoid making this disagreement public.

Shortly after I had the chance to review and comment on PD-42, President Carter made his first speech on the space program at the Kennedy Space Center on October 1, 1978. The speech was essentially a statement summarizing PD-42 and publicly reaffirming the administration's support for the space shuttle program. In achieving this public support from the president, I felt that we had scored a major success and that we would now be able to push the shuttle program through to a successful conclusion. We would then be in a position to build things like the space station on that base when the right time was at hand. In the same speech, the president also made an announcement that was much more controversial. He made public for the first time that the United States operated photoreconnaissance satellites for the purpose of verifying arms control agreements. The fact that this statement was actually included in the speech was remarkable considering the opposition that existed in the intelligence community to any declassification of satellite data and/or operations. The intelligence people were overruled in this case by the president's political advisors. The latter felt, correctly as it turned out, that it would be difficult to secure ratification of the SALT II arms control treaty then being negotiated with the Russians unless we could produce some graphic evidence about the capability our space systems had to find out what the Russians were doing. The entire debate over the verification of SALT II treaty provisions hinged on this point. The president's public announcement was, of course, an essential precursor to securing ratification by the U.S. Senate.

I was not consulted on this point of the president's speech, but had I been, I would have supported an even stronger position on declassification. The president declassified only *the fact* that we were taking pictures, whereas I would probably want to declassify some of the pictures as well. My reason would be somewhat different from the one that motivated President Carter. I had long felt that we should provide more information to the American people about what the Russians were doing in terms of building up their nuclear strategic forces in order to generate more public support for our own strategic force modernization program. October 1, 1978, happened to be President Carter's fifty-fourth birthday and, by coincidence, the then projected date for the first space flight of the shuttle was scheduled for October 1, 1979. Bob Frosch, in his remarks that concluded the ceremony at the Kennedy Space Center, promised the president that the first successful flight of the shuttle would be a fifty-fifth birthday present. Unfortunately, Frosch, as well as the rest of us at the time, were too optimistic, and there would be painful delays before the first successful flight would actually be executed on April 12, 1981. (By coincidence, this was the twentieth anniversary of Yuri Gagarin's first orbital space flight.)

Having now strayed into the areas of arms control and defense policy, I want to continue with a somewhat more extensive discussion of the arms control position during the Carter administration because arms control verification was one of the principal reasons why many senior officials in the administration strongly supported the space program.

One of the top priorities of the Carter administration was to try and reach further agreements with the Russians on the control and the limitation of nuclear weapons. Ever since my service at the Livermore Laboratory, I had been very skeptical about the value of nuclear arms control agreements with the Russians. Except for treaties limiting the actual deployment of existing nuclear weapons, I felt that such agreements would not generally be of value to the United States. I also felt that the Russians tended to look at the world from a fundamentally different viewpoint than ours, and they therefore regarded agreements of this kind rather differently than we did. On the other hand, I also recognized that the U.S. government was (and still is, for that matter) under considerable international pressure to conclude nuclear arms control agreements with the Russians and that it would probably hurt our standing with our friends and allies abroad if we did not make the effort to negotiate.

Since the verification of the existing nuclear arms control agreements was being accomplished largely by the "national technical means" (that is, by the space systems) for which I had overall responsibility, I made it a point to initiate discussions with the people responsible for arms control. On September 22, 1977, I met with Mr. Paul Warnke, the new director of the Arms Control and Disarmament Agency (ACDA), and his deputy, Mr. Spurgeon Keeney. Subsequently, on October 6, 1977, I had a longer session with Keeney. This discussion was a wide-ranging one during which I described to him the technical prospects as I saw them and what might be done to develop new satellite systems that could be used to verify more comprehensive treaties than presently possible. Keeney, who had served in the Air Force for some years, was thoroughly familiar with the military space program and understood very quickly what could be done. Keeney arranged for more thorough briefings of his staff as well as for Harold Saunders (then assistant secretary of state for intelligence, etc.) and various State Department people. These briefings were held on February 1 and 24, 1978.

Two consequences followed from these discussions. One was that the ACDA and State Department staffs strongly supported the new initiatives in the national security related space program that I was trying to push through because they recognized that these new systems could be very useful in the verification of new arms control agreements. The second was that I inevitably became drawn into the debate over the ratification of the SALT II agreement by the U.S. Senate. The former was all to the good, but I am not at all sure that the effort I spent on SALT II was worthwhile.

President Carter's arms control advisors had completed the detailed negotiations on the SALT II agreement with the Russians in the spring of 1979. On June 17, 1979, the SALT II treaty was signed at the "summit" meeting in Vienna between presidents Carter and Brezhnev. On the whole, I did not think the SALT II treaty was a bad agreement —especially in that the proposed limits on weapons deployments were probably somewhat in our favor. I was more concerned over the outcome of the debate on the limits that were proposed on the testing of strategic weapons. These limits tended to hinder the development of new technology, and since we are better at this than the Russians, I felt that such limits would be to our disadvantage. There was one final point that concerned me even more about the proposed treaty and that was timing. I felt that the conclusion of the treaty at that time would create a sense of euphoria in this country that would make it very difficult to continue to gain the public support necessary for the modernization of our own strategic nuclear forces that was then just getting started. I therefore had very ambivalent feelings about the value of the proposed SALT II agreement. There was, however, one point on which I had no real doubts. After making a careful study of the technical situation, I became convinced that the treaty could be properly verified using the space systems we then had in place. Therefore, I could be helpful to the administration in discussing the technical aspects of the verification problem without taking a strong position on the merits of the treaty itself. I did this as best I could, and held several conversations with influential senators—specifically, Barry Goldwater and John Glenn. I do not know whether I had any real influence because the SALT II treaty never came to a vote. When the Russians invaded Afghanistan late in 1979, the effort to push the treaty through the Senate was abandoned.

Shortly after President Carter was inaugurated, the Russians carried out a successful test of their antisatellite system (ASAT). Their development of means to destroy satellites actually dates back to the late 1960s and by the time they conducted this test, their capabilities in the antisatellite area were well known. The Russian system is called a co-orbiting antisatellite weapon. It destroys a target satellite by launching the killer satellite into a similar orbit (so that the killer co-orbits with the target), and a radar or infrared homing system is then used to guide the killer satellite to the target. Once the killer is close enough to the target, an explosive charge is fired and the target is destroyed by shrapnel. President Carter was sufficiently disturbed by the ability of the Russians to do this that he asked them to sit down with us to negotiate a treaty that would hopefully prohibit the deployment of such weapons.

I was thoroughly opposed to such a negotiation because I could only see an outcome that would be disadvantageous to the United States. I felt that the Russian system was relatively primitive and that means could be developed in fairly short order to protect our own satellites. At the same time, I was certain that in any negotiation, the Russians would construe many operations that we were planning for the shuttle —such as satellite pickup for example—as an antisatellite system test and would therefore seek to limit our shuttle operations. Also, I feared that any treaty would limit our ability to deploy the "direct ascent" antisatellite system we had under development. A direct ascent system is more capable than the Russian weapon because the killer vehicle does not have to co-orbit with the target. Thus, it is much harder for the target to avoid being hit because things happen much more quickly.

The Russians accepted President Carter's offer and, in due course, negotiations were started. Once the negotiations began, I could not oppose our participation, but I tried to influence our negotiating position to minimize the damage that might be caused by limitations on our ability to operate the shuttle or to develop antisatellite systems of our own. Once again, I would not be directly involved in the policy-making process on this issue, but two people would be very helpful in keeping me informed, Mr. Philip E. Culbertson, an old friend from my NASA days, who represented NASA on our ASAT negotiating team, and retired Air Force Major General David Bradburn, who represented the Joint Chiefs of Staff. Through the good offices of Culbertson and Bradburn, I was able to make my views known and to keep informed. There were two fairly lengthy negotiating sessions, which ended essentially in an impasse. As expected, the Russians were anxious to limit the operational capability of the shuttle, something that we refused to do. The Russians, on their part, refused to agree to stop testing their ASAT system. After the Russian invasion of Afghanistan, the negotiations were stopped.

During the two years I served as under secretary of the Air Force, I tried as best I could to influence the Air Force organization for the conduct of space activities. I quickly became convinced that it would eventually be necessary to establish an operational Air Force Space Command in order to properly manage and deploy our military space system. An operational command would also be important for the advocacy of budgets for space operations in the senior councils of the Air Force. I also wanted to establish a development agency that would be exclusively responsible for the development of military satellite systems. I have already mentioned the reorganization of the Air Defense Command that had been proposed by the then Air Force chief of staff, General David C. Jones, and I felt that the changes proposed would provide us with a good opportunity to establish a Space Command as part of a reorganized Air Defense establishment (see chapter 7). My new military assistant, Colonel (now Major General) Thomas W. Sawyer, would be very influential in shaping my thinking on this important matter. (Sawyer succeeded Harry Goodall in this post on April 8, 1978.) Sawyer had spent many years in the Air Defense Command and was thoroughly familiar with the whole problem. He was also a major participant in the reorganization study that General Jones had requested so that he understood the circumstances that had moved the chief of staff to propose the reorganization.

On August 22 and 23, 1978, I again visited General Hill at his headquarters in Colorado Springs to talk about the proposed reorganization of the Air Defense Command. We discussed the idea of establishing a Space Command in more detail and began to evolve a strategy as to how this might be accomplished. In spite of the fact that, as under secretary of the Air Force, I had broad overall responsibility for Air Force space programs, I did not have much leverage in organizational matters. I had a platform from which I could argue and could refuse to concur with what was being proposed, but the chief of staff and the secretary had the final authority. On February 8, 1979, I attended a senior Air Force Commander's Conference at the Homestead Air Force Base near Miami, Florida, and I used the occasion to make a speech advocating the establishment of an Air Force Space Command in connection with the Air Defense Command reorganization. Although my proposal was rejected, I did receive some unexpected support from some of the senior officers present—especially from Lt. General Andrew Anderson, who was then serving as deputy chief of staff of the Air Force for operations. A week later, in Washington, I met with both John Stetson and the chief of staff (General Lew Allen, Jr., who had succeeded General Jones when the latter became chairman of the Joint Chiefs of Staff in July 1978). It was made clear to me at this meeting that the Air Force would go ahead with the proposed reorganization of the Air Defense Command, in spite of my objections, and a few weeks later, John Stetson signed the final approval. It is ironic that this step was taken three months before Stetson would himself resign as secretary of the Air Force and I would be asked to succeed him. Perhaps a Space Command would have been organized more rapidly had I been secretary at the time, but, in hindsight, I doubt it. The conditions were not yet right in the spring of 1979.

I had more luck with the establishment of an organization to take exclusive responsibility for the development of Air Force Space Systems. On July 27, 1977, my old acquaintance (from Project Argus days) General Lew Allen, Jr., replaced General William Evans as the commander of the U.S. Air Force Systems Command. In that post, he would oversee all of the technical development programs of the Air Force. Furthermore, Allen was completely familiar with Air Force Space Programs having served in a number of senior positions with responsibility for them. In addition, Allen had (and has) a very acute mind, first-class technical judgment, and a broad viewpoint on what is important. Although I did not know it at the time I attended the change-of-command ceremony, this event would mark the beginning of a very fruitful and pleasant association with Allen that spanned my years in Washington, both in the Air Force and in NASA. Allen and I had a number of meetings after his appointment to head the Air Force Systems Command. It quickly became apparent that we were in general agreement on most of the organizational problems we were facing and that Allen also supported the establishment of an organization with exclusive responsibility for the development of Air Force Space Systems. As things turned out, Allen would not remain in his new post for very long. He was appointed vice chief of staff of the Air Force in April 1978 and chief of staff in July 1978. It fell to Allen's successor at

Systems Command, General Alton D. Slay, to implement the plans that Allen and I had discussed during our meetings in late 1977 and early 1978.

On April 16, 1979, John Stetson informed me that he would be leaving his post as secretary of the Air Force on May 19. About two weeks later (May 2), Harold Brown asked me to succeed Stetson. Initially, I had some misgivings because my job as under secretary permitted me to exert considerable influence on the management of the nation's space program. I knew that if I accepted Brown's offer, I would have to give that up. On the other hand, the opportunity to learn something about the broader issues related to our national security was enough to override my doubts. In addition, as secretary, I would have greater influence on the development of the appropriate organizations within the Air Force concerned with the development and operation of space systems. Accordingly, I accepted Brown's offer. Even though all this occurred a year after my initial conversations with Allen and Slay on organizational issues in Systems Command, my elevation to the job of secretary of the Air Force gave me the chance to do something quickly and decisively to improve the organization of the space activities and space operations of the Air Force. Shortly after I was sworn in, I went to Los Angeles to announce the establishment of the Air Force Space Division. In a speech before the local chapter of the Air Force Association on August 3, 1979, I announced that the Air Force Space Division would be established as one of the major "product" divisions of the Air Force Systems Command. The opportunity to take this step presented itself when the MX program to modernize our land-based strategic missile force went into high gear. In view of this, it was decided to split the then existing Space and Missiles Systems Organization (SAMSO) of the Air Force Systems Command into two components, one to manage the MX strategic missile program, which would be located at Norton Air Force Base near Riverside, California, and the other to manage the development of space systems. The new Space Division would be located at the Los Angeles Air Force Station and would be the first major Air Force unit commanded by a three-star officer devoted exclusively to the development of space systems. This was an important step but it was only one step because the operational responsibility for space systems was still divided between several major commands. This situation was rectified by my very distinguished successor as secretary of the Air Force, the Honorable Verne Orr, when he authorized the establishment of the Air Force Space Command in September 1983. This step was the implementation of the plan I had discussed with General Hill in 1978 (see chapter 7), and it called for the creation of a Space Command colocated with the North American Air Defense Command in Colorado Springs, Colorado. General James V. Hartinger was inaugurated as the first commander of the Air Force Space Command. Needless to say, I was most gratified by the eventual outcome.

I have dwelt on these organizational issues because I believe that organizations tend to shape people's thinking. I felt that it would be easier for the Air Force to accept some of the new things that were being done in space if appropriate organizations were in place to manage them, both in the technical and the operational organizations. There is no doubt in my mind that both the space shuttle and the space station will have important roles to play in our national security related space programs. I believed then, and I still believe, that both the Air Force Space Command and the Space Division are important elements in permitting the Air Force to take advantage of the new technology that is being developed. Whether this judgment was (and is) correct, only time will tell.

When I became the secretary of the Air Force, I asked Antonia Handler Chayes to succeed me as under secretary, and Dr. Robert J. Hermann to become assistant secretary for research, development and logistics, succeeding Dr. John J. Martin, who had recently resigned. Hermann would also eventually take over some of the responsibility for the Air Force Space Programs that is normally assigned to the under secretary. The three of us were sworn in on July 26, 1979. Chayes had turned in an effective performance as assistant secretary for manpower and reserve affairs. Although I disagreed with her on many issues, she eventually developed into a very valuable deputy for me. She took over the difficult job of conducting the advocacy for the MX program, and she provided a political pipeline for me into the senior levels of the Carter administration that I would not otherwise have had. Hermann was a professional engineer and civil servant of great ability. He had spent over twenty years serving as an engineer, a manager, and, finally, as an assistant director of the National Security Agency. After that, he had worked in the Office of the Secretary of Defense.

Hermann was thoroughly familiar with the national security related space programs and, during his short term of service, would make a strong and effective contribution. The three of us would serve in our respective posts for the remaining eighteen months of the Carter administration.

IX The Shuttle Program Has Problems

The first space shuttle flight vehicle, *Columbia* (Orbiter Vehicle-102), was rolled out of its hangar at Palmdale, California, on March 24, 1979. This was a red-letter day for those of us who were interested in the shuttle program and who wanted to see it succeed. Unfortunately, we also knew that there would be real problems to solve with respect to maintaining the projected flight schedule. As early as November 22, 1978, at a meeting I attended to discuss NASA-Air Force space collaboration, both Dr. Alan M. Lovelace (NASA deputy administrator) and Mr. John F. Yardley (NASA associate administrator for manned space flight) hinted that there were technical problems that might make the then-projected late 1979 first launch (or First Manned Orbital Flight—FMOF) impossible. But much worse was to come. Shortly after *Columbia* was rolled out, it was discovered that many of the "tiles" that constitute the shuttle's Reusable Surface Insulation (RSI) system could easily be pulled away from the structure of the shuttle. The worst fears that Howard Larson and his friends had expressed when I had written my letter to Roy Jackson on the "technology readiness" of the shuttle program seven years earlier would be realized (see appendix 1). Just as they had feared, not enough attention had been paid to the adhesive that was used to bond the tiles to the structure of the shuttle and to the development of test procedures to make certain that the adhesive actually worked. All of this would now have to be fixed under the most difficult circumstances and under the glare of unfavorable publicity.

There were really two separate technical problems with the tile system as it stood. One was that the adhesive was faulty and that the aerodynamic shear forces that were predicted during space flight were large enough to make some of the tiles come off. The second was that some types of tiles were weak enough so that the stronger shear forces that would be experienced during the launch ascent trajectory of the shuttle through the atmosphere would cause them to break off not along the bond line to the aircraft structure but at a point in the body of the tile itself. This latter phenomenon had been discovered once reliable "pull tests" had been devised to test whether the tiles would come off under certain conditions. Thus, at least two failure modes were discovered that could cause tiles to be lost or damaged in such a way that the shuttle's structure might overheat during the reentry maneuver. Much effort went into finding the solutions of these problems—and they were eventually solved. However, their existence caused very painful delays and cost overruns that made the years 1979 and 1980 the most difficult ones of the shuttle development program.

In addition to the problems with the tiles of the Reusable Surface Insulation System, there were also technical problems with Space Shuttle Main Engines (SSMEs). The SSME is by far the most advanced rocket propulsion system ever built. It operates at a combustion chamber pressure about four times as high as any previous liquid hydrogen-liquid oxygen rocket engine. Thus, there were much greater power densities involved. Also, the engine was designed to fly fifty times; that is, it also had to be reusable just as the tiles of the thermal protection system were reusable. As things turned out, there were numerous failures of Shuttle Main Engines on the test stand in 1978, 1979, and 1980. Most of these were caused by the very high vibration levels and the very severe thermal and mechanical stresses that some components of the Shuttle Main Engine were subjected to during operation. There were bearing and seal failures in the turbomachinery, turbine blades cracked for various reasons, there were metal fatigue failures in valve bodies and pipes, as well as numerous other problems. All of these problems had to be discovered, diagnosed, and fixed. That all of these problems were finally understood and solved is a major technical triumph much to the credit of the people who brought it about.

There was one final technical problem that had to be dealt with that was less acute—only perhaps because we knew less about it—than the others. The shuttle on its first flight would fly a trajectory that passed through all the flight regimes—space, entry, hypersonic, supersonic, transonic, and subsonic—something that was completely unprecedented. The flight controls and the control computers had to work correctly, and they had to work the first time. Unfortunately, the control laws we employed to program the flight computers we knew were based on necessarily inadequate wind tunnel test data because we did not have the test capability to cover the entire regime through which the shuttle would fly. Nor were our theoretical techniques good enough to make calculations that could be relied upon with the degree of certainty that was normal, say, in the design of a conventional aircraft.

It is important to recognize that at least some of the problems we were now facing were due to the fact that the entire shuttle program from its inception in 1972 to its completion was funded at a level that was chronically below what had been promised by President Nixon (and the OMB). When the shuttle program was started early in 1972, NASA Administrator James Fletcher had made an agreement with the president and OMB that the shuttle would have to be developed within a constant NASA budget of about $3.4 billion (1972 dollars). Had this bargain been maintained, I believe that the problems that beset us in 1979 and 1980 could have been avoided because we could have conducted the kind of test programs that would probably have allowed us to solve the problems we encountered earlier in the program. As it was, the lack of money in the early and mid-1970s caused corners to be cut and risks to be taken, and the resulting difficulties would come home to roost in due course. What happened was that the inflation rates that the nation suffered during the Ford and the Carter administrations were much higher than anyone had anticipated. In view of this, neither Fletcher nor his successor, Robert Frosch, were able to persuade the White House to live up to the bargain of 1972.

In the last half of 1979, there were numerous meetings to determine how high the cost of the shuttle program would actually go. On August 14, 1979, I met with Mr. David Williamson who informed me that NASA would need at least $500 million more in Fiscal Years 1981–85 to complete the shuttle program as it was then laid out. (Williamson was at that time serving as a special assistant to the NASA administrator, a post he held for a number of years and in which he rendered important and distinguished service to NASA and to the nation.) A day later, I called the NASA comptroller, Mr. William E. Lilly, who was an old and trusted friend from my days at Ames. He essentially confirmed Williamson's story and added that NASA would also need about $150 million extra in Fiscal Year 1980—the fiscal year that would start on October 1, 1979—or, in only six weeks. To do this would mean that NASA would be forced to submit a supplemental budget request to the Congress. This was something that Lilly especially was very reluctant to do since it was a source of great pride to him that NASA had always conducted its financial affairs in such a way that supplemental actions were not required. This would be a break in the tradition that Lilly had established.

A few weeks later, I visited NASA headquarters to have lunch with Frosch, Lovelace, and Lilly (August 30, 1979), and by now the figure had grown to $200 million in Fiscal Year 1980 and $300 million in Fiscal Year 1981. They also told me at this meeting that they were really not yet certain of the precise figure. They did know that it would be a large number. Finally, on September 5, 1979, Max Faget came by and gave me a report on the technical situation as he saw it. Faget felt that the cost overrun we would see in the shuttle development program would exceed the $500 million estimate that I had heard. Furthermore, he felt that the date of the First Manned Orbital Flight would be delayed by at least a year. The fact that all of these estimates were somewhat different indicated to me that the full extent of the cost problem was not really understood. As things turned out, the ultimate cost overrun was well over $1.0 billion in 1972 dollars and the first flight originally scheduled for 1979 was delayed ultimately to early 1981.

All of this was really bad news. Not only would the shuttle itself be late and more costly, but the remainder of the system would suffer as well. I have already described in chapter 5 how the work on the whole shuttle program was divided between NASA and the Air Force. Once the NASA part of the program was in trouble, it was clear (and logical up to a point) that the Air Force should back off and delay the completion date of the west coast shuttle launch facility for which it had the responsibility. The trouble was that there were (and still are today) some people in the Air Force who argued that the whole commitment to the space shuttle program was a mistake and that the entire issue should once again be critically reviewed. This was done in the late summer and early fall of 1979 and, once again, the old dispute surfaced.

Since I was by now secretary of the Air Force, my responsibilities were much broader than in my previous post. I had to think in terms of the entire Air Force mission and not only about the space portion of the Air Force program. Early in July 1979, I drafted a statement of priorities for the Air Force that I would try to translate into programs that could be implemented. It laid out three priorities along with their programmatic consequences:

1. *The modernization of our nuclear strategic deterrent forces.* I felt this was clearly the first priority because we had to do something to counter the massive buildup of the Russian nuclear strategic forces that they had initiated after the Cuban missile crisis in 1962. In terms of programs, this meant the completion of the new MX strategic missile and the development of a new manned bomber. The MX was not controversial at the time because it was supported by the Carter administration. This was not the case for the bomber. It was controversial because I advocated the revival of the B-1 bomber program in some form rather than the alternatives that were favored by uniformed Air Force people (modifying some FB-111s to give them more range payload capability) and by Bill Perry and Harold Brown (building a fleet of airplanes based on "stealth" technology). I felt that the modified FB-111s would not be capable enough to do the required job and that a deployed force of "stealth" bombers would not be available until the mid-to-late 1990s given the technical difficulties that were bound to be encountered. I therefore advocated reviving the B-1 program (I used the euphemism "Long Range Combat Aircraft" to do this), which unfortunately strained my relations both with Perry and Brown. After the Russians invaded Afghanistan in December 1979, I felt particularly strongly that an attempt should be made to persuade President Carter to reverse his decision in 1977 to suspend the B-1 production program. I failed in this attempt and the B-1 would not be revived until the Reagan administration took power in 1981.

2. *The enhancement of our airlift capability.* Given the responsibilities we had assumed in the world, I thought that our ability to rapidly deploy our military forces was dangerously weak. We had capable aircraft in the C-5 and the C-141 for long-haul air transportation although we did not have enough of the airplanes in either category and the C-5s were flawed because of the relatively short fatigue-failure lifetime of the wings. We also had a large fleet of very good tactical transports, the C-130s, but many of these aircraft had seen hard service in Vietnam and were beginning to reach the end of their useful lives.

We were able to push along and formally initiate the program to stretch (i.e., lengthen) the fuselage of the C-141 aircraft, thus enhancing the payload weight and volume capacity by a substantial factor for a relatively small cost. We also succeeded in pushing through the program to fix the wings of the C-5s. This was done in spite of strong opposition from Senator William Proxmire and his staff that culminated in a contentious hearing on September 16, 1980, with the senator in the chair. We succeeded in maintaining the C-5 wing modification program. We failed to initiate a program to replace the C-130s, and we were unable to increase the number of airplanes in our inventory. Both of these objectives were eventually accomplished under the Reagan administration with the initiation of the C-17 tactical short-take-off-and-landing (STOL) air transport program and with the purchase of fifty additional C-5 aircraft.

3. *The improvement of our ability to conduct operations in space.* My two years as under secretary had convinced me that space operations for national security related purposes were becoming much more important than they had been in the past. I have already mentioned the problem of verifying arms control agreements as something that depends heavily on satellite systems. There are other applications as well including the support of ground and naval forces through the provision of timely, combat-related information. There is the very important area called "Indication and Warning" where satellites provide much of the critical information that—in the case of strategic nuclear attacks —provides warning and also characterizes the size and the composition of the attack. Finally, there is the possibility of eventually developing a system to defend against Inter-Continental Ballistic Missile (ICBM) attacks and Submarine Launched Ballistic Missile (SLBM) attacks. Any such system would depend very heavily on enhanced operations in space at least for the provision of data and perhaps for the deployment of actual weapons systems. (Late in February 1980, I wrote Harold Brown a long personal note on MX in which I argued that deception schemes for assuring the survivability of the MX would either not work or would be politically unacceptable. I urged that we consider putting the MX missiles in fixed silos and then defending these silos with an appropriate antiballistic missile system. I pointed out that Brown's continuing strong support of the space program would make this a natural position for him to take. My advice was not accepted.) The most

important programmatic priority I put in this category was that the Air Force must live up to its commitment to support the space shuttle program. This meant completing the west coast shuttle launch site at the Vandenberg Air Force Base and learning how to use the new capabilities that the shuttle represented to the best advantage for our national security. As things turned out, this programmatic proposal was, and still is, somewhat controversial.

On September 19, 1979, I was able to use the forum of the annual Air Force Association meeting in Washington to make a major speech outlining these priorities and giving the reasons why I believed they were important. The speech was reprinted in the September 1979 issue of the Association's Magazine *Air Force* (appendix 3). I have digressed somewhat from describing the problems of the space shuttle program because in my new post as secretary of the Air Force I could not devote nearly as much time to the space programs as I did when I was under secretary. There was, and still is, no doubt in my mind that both the strategic force modernization and the airlift enhancement were more important at that time to our national security than the space program, and I divided my time accordingly. In spite of the fact that I could not spend nearly as much time on the space related problems as I had in the past, being secretary of the Air Force did have some advantages. One was that I had the final sign-off authority on the Air Force budget and could therefore exert important leverage. The second was that I now belonged to the PRC (Space), which gave me direct access, for the first time, to the policy-making apparatus in the White House. The second point turned out to be completely unimportant because by the time I joined the PRC (Space), all the important policy documents had already been written. In my eighteen months as a member, I went to only one meeting (on October 4, 1979), and nothing of substance came out of it.

I did do one thing at this time that did have lasting value by initiating the "Space Breakfasts" at 7:00 A.M. in my office every Friday. The membership of the group included senior NASA and Air Force officials (both civilian and military) involved in space matters and the "Corn Flakes Club," as it soon became known, was an excellent forum for discussion. When I left the Air Force in 1981 and rejoined NASA, the 7:00 A.M. time was abandoned, but the meetings were reconstituted as the NASA/DOD lunches every second Tuesday. Naturally, these were soon called the "Corned Beef Club."

The central issue that had to be faced at this time was the development of the proper role for the Air Force in the space shuttle program. While the 1972 agreement defined a clear role for the Air Force in the shuttle development program, the fact is that most of the leaders of space activities in the Air Force were not happy with that role. In short, the Pentagon's civilian leadership over the years favored the shuttle program and supported the Air Force role as a matter of national policy. However, those people who had to execute the program in the field were not nearly as favorably inclined. There were two major issues that were not resolved during the years that I served in the Pentagon and that are really still not resolved today:

1. What is the value of people in space for military operations?

2. How can the Air Force execute the appropriate control over a space launch vehicle (the shuttle) that it does not operate?

These are both legitimate questions, but it is important to recognize that there are no simple answers. The study that Tom Stafford and I headed on the uses of military people in space was perhaps not persuasive enough. Furthermore, the experience that the Air Force had with the Manned Orbiting Laboratory (MOL) program had left scars. In addition to the usual reasons for not using people in space (expense, vulnerability, etc.), there was also the residual feeling that nothing good had come out of previous Air Force attempts at promoting a manned space flight program and that any new efforts would wind up the same way. I had numerous discussions with many people in the Air Force about the question of manned space flight during the four years that I served in the Pentagon, and I failed to make any significant headway in persuading people of the military value of manned space flight. The difficulty, of course, was that the central point of my argument is not subject to analysis—rather, it is an article of faith: people are valuable anywhere because of the elements of judgment and imagination that they can bring to dealing with situations they face. Since space flight and space operations have important national security related functions, the presence of military people in space should be of value. Therefore, it is worthwhile to spend a reasonable amount of extra money to put the people where they are needed. This is, of course, exactly the argument that the Air Force uses to justify manned bombers, for instance, but as yet the connection in the case of space operations is apparently not strong enough to persuade most military people interested in space operations.

The budgetary issue over which this particular argument came to a head was the construction of the west coast shuttle launch site at Vandenberg Air Force Base. This job is an extremely complex engineering construction problem, and as happens in so many of these projects, problems that were not anticipated by the project management had developed. Some of these problems were caused by a faulty relationship between NASA and the Air Force at the time, which did not give the Air Force enough management freedom to do the job properly. In any event, these problems led to large cost overruns and schedule delays. I reviewed the situation numerous times in the summer and fall of 1979, and later on in 1980 for the Fiscal Year 1982 budget, and came to the conclusion that the cost overruns could be handled within the Air Force budget. This was not to be the case for NASA and the cost overrun problem on the shuttle Orbiter program. The total NASA budget was only one tenth of the Air Force budget and, thus, their overrun was a very much larger fraction of their budget. Thus, the NASA budget problem became much more of an issue—a matter that I will discuss shortly.

There were many senior members of the Air Staff in Washington and also senior officers in the field who were not in favor of doing the space shuttle program in the first place. Accordingly, when the problems at Vandenberg became apparent, great pressure developed to cancel, "suspend," or delay for a long period the construction of the shuttle launch site on the west coast. I was strongly opposed to these suggestions. I believed then, and I am even more convinced now, four years later, that people in space will ultimately perform very important military functions and that the capability for manned flight of the shuttle will be the critical element in permitting them to do this. Furthermore, I felt that the shuttle program that we agreed to at the beginning of the Carter administration—four Orbiters and two launch sites—was the minimum possible program that would permit us to explore the full utilization of the shuttle and any future steps, such as a permanently manned space station, would be completely impossible to take without the prior execution of the minimum shuttle program.

In view of all this, I exercised my prerogative as secretary and refused to sign out any Air Force budget requests that did not include full funding for the shuttle launch facility at Vandenberg. I did this for the two budget years (Fiscal Years 1981 and 1982) when I had the authority, and the Air Force budget for those years provided the construction money for the project. The Vandenberg shuttle launch facility thus became my "Gold Watch," which is Pentagon jargon for any program that is imposed on the military establishment by a member of the civilian hierarchy. This is not the first time—and certainly not the last—that civilian judgment has been part of the preparation of the military budget, and it is, in fact, how our military system is supposed to function. I know that these decisions on my part caused some hard feelings, and I tried to minimize the adverse effects by fighting even harder for those programs that the military wanted elsewhere in the Air Force budget.

Even more important than the question of what people could do in space is the matter of who controls the launch vehicles. This was, and still is, a primary motivating factor in the thinking of many senior Air Force people both military and civilian. The fact that the shuttle was not under management control of the users concerned many people. Specifically, they were worried about NASA's ability to meet Air Force schedules and particularly to carry out some launch-on-demand type missions that the Air Force might want to execute. There was not much I could do directly about this as long as the shuttle was (and is) operated by NASA. What I could do (and did) was to develop the idea of "dedicated" Air Force missions in which the Air Force was in complete control of the mission even though NASA actually owned the vehicle. I thought that if the Air Force had a facility from which it could actually perform the mission control function for the shuttle, then perhaps some of the objection to using the shuttle would be blunted.

It turned out also that I had some leverage in this case that I could not muster in the case of the Vandenberg facility. For a number of years, the Air Force had made plans to move their major ground facility at Sunnyvale, California—the Satellite Test Center, or STC—to a more secure location. They also wanted to expand the capability of the facility, which would be called STC-II or "Stick Two." I was not a strong supporter of this proposal, but I was able to strike a bargain with the leadership of the Air Force Systems Command and its Space Division (which was by then in existence). I would support "Stick Two" if the facility included not only satellite control facilities but also a mission control center for the control of shuttle flights. Thus, the concept of the Consolidated Space Operations Center (csoc) was born. The dispute in the case of the csoc was not over whether it should be done at all, but where it should be located. I asked that the Air Force conduct a site study for csoc, and the recommendation was that the facility be located either at Malmstrom AFB in Montana, at Peterson AFB in Colorado, or at Kirtland AFB in New Mexico. Malmstrom was quickly eliminated on the ground that proximity to a well-developed industrial community would be important. This left Kirtland and Peterson as possible locations. I strongly favored a location near Colorado Springs for the csoc because I had not given up on the idea of eventually establishing an Air Force Space Command based on the Air Defense Command. I felt that the location of csoc near Colorado Springs would be an important step in that direction.

A spirited competition now developed between those favoring the New Mexico site and those who wanted to see the new facility in Colorado Springs. And, as is usual in cases such as this, the congressional delegations from the respective areas soon became involved. Within the Air Force there were also two camps, and it took some months of debate and argument to resolve the issue. The matter was settled on October 27, 1979, when the Air Council (an air staff Committee consisting of all the deputy chiefs of staff of the Air Force chaired by the vice chief of staff) approved the Colorado Springs location, and so the arena of the debate shifted to the Congress. The two principal advocates for the different locations were senators Harrison Schmitt (R., New Mexico) and Gary Hart (D., Colorado). It was unfortunate that I could not agree with Schmitt on this question since it strained what had long (since 1969) been a good relationship. Hart, of course, favored the proposal and his support—and the very effective staff work by his assistant, Mr. Larry K. Smith—turned out to be critical not only for csoc but for the construction of the west coast launch site at Vandenberg Air Force Base as well.

At the time (1979), Hart not only represented the interests in Colorado but was also serving as chairman of the Military Construction Subcommittee of the Senate Armed Services Committee. In that position, he had a very strong influence not only on csoc but also on the authorization of the west coast shuttle launch site. I had numerous conversations with Hart to explain to him the importance of our space program to the national security. Hart, who is extremely perceptive and knowledgeable, quickly understood what might be done and became a strong supporter of both CSOC and the Vandenberg facility. He spearheaded the effort in the Senate to push through both of these facilities. Hart's opposite number on the Appropriations Committee, Senator Walter Huddleston (D., Kentucky) and his chief staff person, Ms. Carolyn Fuller, were equally effective and important in developing and maintaining support for these facilities. In the House, representatives Gunn McKay (D., Utah) and Lucian Nedzi (D., Michigan) were the individuals who proved to be the most important in maintaining the construction program that we had proposed.

The problems related to covering the cost overruns in the Vandenberg construction budget and the development of the CSOC could be handled within the Air Force budget without involving the White House and the OMB to request large supplemental appropriations. Unfortunately, this was not true in the case of NASA, and a strong and conscious effort had to be mounted to secure the necessary funding for NASA to maintain some reasonable schedule for the completion of the shuttle development program. Ever since I arrived in the Pentagon in 1977, I did my best to persuade people that the space shuttle would eventually have extremely important applications related to national security. Both Harold Brown and Bill Perry became strong supporters of the space shuttle program on the ground that the new capabilities that the shuttle would provide would eventually be important to the national security. Brown had a good opportunity to make the case publicly when he testified before the Senate Subcommittee on Science Technology and Space of the Committee on Commerce, Science and Transportation then chaired by Senator Adlai E. Stevenson III. Both Brown and I were witnesses at the hearing held on February 7, 1980, during which we made strong statements supporting the NASA request for funds to deal with the cost overrun problem and to keep the shuttle program on schedule because it would be important for national security.

It was, of course, the same argument, that the shuttle would be important for the purposes of national security, that persuaded President Carter to support NASA's supplemental budget request and to send it to the Congress in the first place. Here again, the roles played by

The meeting to continue the shuttle program. From left: Frank Press, James McIntyre, President Carter, Robert Frosch, Mark, Alan Lovelace, and W. Bowman Cutter.

Brown and Perry were critical. The final meeting confirming the decision by President Carter to continue his support of the shuttle program in spite of the difficulties NASA was facing was held in the Cabinet Room in the White House on November 14, 1979. It was a fairly large meeting with Jim McIntyre and his deputy, W. Bowman Cutter, speaking for OMB; Frank Press and Ben Huberman for the Office of Science and Technology Policy; Bob Frosch, Alan Lovelace, John Yardley, and Bill Lilly were there representing NASA; and I spoke for the Department of Defense and the Air Force.

The president opened the meeting with a statement that he had decided to support the shuttle program and that he wanted some kind of statement for the administration to that effect. (This decision—in its final form—had apparently been reached earlier that day in a private meeting between Frosch and the president that was held in the Oval Office.) The president went on to stress the importance of national security related applications of the shuttle. He then asked NASA for a report on the technical status of the shuttle program and told us what he wanted to come out of the meeting. First, he said he wanted to confirm his decision to support the shuttle program. Second, he said that he wanted to designate a single "point-of-contact" in the White House to deal with matters related to the shuttle program. And, third, he wanted a technical status report from NASA. Bob Frosch proceeded with the status report and discussed both the technical and financial problems the program had encountered. The president asked a number of detailed technical questions about the problems and the delays in the engine test program and with the tiles of the Reusable Surface Insulation System. It was quite obvious that he was trained as an engineer! Frosch also discussed the mission model and once again stressed the importance of the flights the shuttle would be conducting for the Department of Defense. During the financial part of Frosch's talk, he predicted that the cost overrun in the shuttle development program would be about 20 percent of the total runout cost or a little over one billion dollars in 1972 dollars. Frosch also estimated that the First Manned Orbital Flight (FMOF) of the shuttle would take place late in the summer of 1980. When Frosch was finished, I made a short statement of support and discussed some of the things that could be done with the shuttle that would be important for national security related space programs. The final speaker was Frank Press, who also made a strong statement supporting the completion of the shuttle program.

At the end of the meeting, President Carter asked Bob Frosch to develop a strong statement of support for the shuttle program in the president's name. He also designated Jim McIntyre, the director of OMB, as the single "point-of-contact" for shuttle related matters in the White House. Finally, he requested that Frosch keep both him and Vice President Mondale informed of the plans for the first flight. He said that they both wanted to know the names of the astronauts and be aware of the activities that would be connected with the first launch and landing. There was no doubt in my mind as I listened to the last part of the president's statement that he was completely aware of the public attention that the first shuttle flight would attract and that this public attention would be politically important.

From this time on, there was generally strong support for the shuttle program from the White House staff as we prepared to support NASA's supplemental budget request before the Congress. Also, Dr. John Koehler who was then serving as the deputy director of Central Intelligence for Resources, became a very strong and effective supporter of the space shuttle program. Therefore, when the time came to present our case to the Congress there would be a strong and united front. In preparation for the congressional hearings that would now be held to consider NASA's supplemental budget request, I held a number of technical and financial reviews in order to convince myself that the program was sound. The most important of these was held on Saturday, March 1, 1980, at the Rockwell International Corporation facility in Downey, California, where the major portion of the production work on the shuttle was being performed. At that meeting, I became convinced that the technical problems were on the way to being solved but that we still had to establish the criteria that would allow us to be comfortable with the risks that would have to be taken the first time that the shuttle would be launched. Accordingly, a week later (Saturday, March 8, 1980), I convened a meeting at the Ames Research Center to convince myself that the technical risks, as best as we could estimate them at the time, were indeed acceptable. I had to do this in order to make certain that I could testify on the status of the shuttle program with complete confidence when the time came to go to the Congress.

The meeting at Ames was attended by senior officials of the Space Shuttle Orbiter Project Office at the Johnson Space Center led by Dr. Milton A. Silveira, the deputy project manager, and by the senior technical people at Ames with center director, Mr. C. A. Syvertson, leading the group. We had a long and wide ranging discussion on the most critical problems we were facing in the main engine program and also in the Reusable Surface Insulation (RSI) program to assure ourselves that the thermal protection system (that is, the famous "tiles") would actually work. In the case of the engines, we concluded that the engines could be operated at an acceptable risk level at the so-called "Rated Power Level," although they were not yet ready for operation at "Full Power Level," which is 9 percent above "Rated Power Level." Since no real payload would be carried on the first flight, operation at "Rated Power Level" was all that would be required, so we judged that the engines were ready to go. The situation with the thermal protection system (or Reusable Thermal Insulation System) was more complex. Several of the technical people at Ames felt that more testing would be required before we could safely go ahead with the first flight. After a long and exhaustive discussion of this issue, I also finally concluded that the thermal protection system could be flown at an acceptable risk level under the conditions planned for the first flight. There was also some discussion at this meeting of the status of the Flight Control System. While there were some problems in this area, I finally concluded that the really first class pilots that would be assigned to fly the first mission could deal with them. The upshot of these meetings was that I was now convinced that the risks we would have to take to fly the shuttle for the first time were indeed acceptable and that I could testify to this point before the Congress in good conscience.

I have already mentioned the hearing before the Senate Space Subcommittee on February 7, 1980, at which Harold Brown and I were the principal witnesses. On February 11, 1980, a similar hearing was held before the House Space Subcommittee chaired by Representative Don Fuqua (D., Florida). I spoke for the Defense Department at this hearing and once again made a strong statement of support for the shuttle program. By far, the most important of the congressional hearings on NASA's supplemental appropriation request to keep the shuttle program on track was held by the House Appropriations Committee on March 13, 1980. This was the meeting I had prepared for in such great detail. It was a joint hearing of the NASA Appropriations Subcommittee chaired by Representative Edward Boland (D., Massachusetts) and the Defense Appropriations Subcommittee headed by Representative Joseph Addabo (D., New York). Senior officials, both of NASA and the Department of Defense, made the strongest possible case that the shuttle program should be continued on schedule and that funds for the NASA supplemental request should be approved. Our arguments were apparently persuasive and Congress promptly appropriated the needed funds. In addition we had strong help in the Senate from Charles McC. Mathias and Harrison Schmitt, who muscled our request past their chairman, William Proxmire. (Proxmire had been a longtime opponent of the shuttle program.)

Another very important effort that was underway at this time was the negotiation of a new Memorandum of Understanding between NASA and the Air Force that would define the relationship between NASA and the Air Force once the shuttle became operational. Mr. Philip E. Culbertson was the senior NASA official in these negotiations and the

Air Force was represented by Major General James Brickel. The essential problem was this: if the shuttle would ultimately become the sole means of getting things into space—and that was the plan at the time—then the Air Force would have to have the right of absolute priority in the launch operations. What we insisted was necessary in view of the vital importance of certain space satellite systems to the national security was an "assured access to space." This meant that under certain circumstances, the Air Force would have the right to take all other payloads (commercial, scientific, and nonmilitary governmental) off the shuttle in order to fly things required by national security considerations. Needless to say, this was a controversial position and the negotiations took a number of months to complete. Several times during these talks, I met with Al Lovelace and Bob Frosch, and both expressed the concern that by taking the position I did, I would cause the "militarization" of NASA. I did not agree with their arguments since I believed (and still believe) that the Defense Department would be a customer like everyone else and that the independence of NASA would not be compromised. I also pointed out to my NASA friends that the support we had provided to NASA in the matter of the supplemental appropriation required that NASA yield on this point; otherwise, I would not be able to resist the pressure from within the Air Force to retain a launch vehicle production capability independent of NASA. I finally won the argument, and the Memorandum of Understanding was signed by all concerned on February 25, 1980.

During the spring and summer of 1980, it became apparent that a number of people connected with the NASA committees in the Congress were increasingly unhappy with the conservative policy with respect to space activities that had been adopted by the Carter administration three years earlier. This policy called for a "balanced" program of space science, space applications, and space technology development, and it proscribed any large new engineering initiatives on the scale of the Apollo program. As I have already said (see chapter 8), I was unhappy with such a negative statement, but I could not change it in 1978. Now things were different. The shuttle program was nearing the end of its development phase, and the time had indeed come to think about what to do next. I was completely convinced that our thinking back in 1969 was sound and that the next step would be to go ahead with the construction of a permanently manned orbiting space station once the shuttle development program was completed. Bob Frosch was somewhat of the same mind, although he was less sure that the space station was the next logical step. In any event, things came to a head when Chairman Don Fuqua (D., Florida) of the House Science and Technology Committee scheduled hearings on the administration's space policy on July 24, 1980. There were three principal witnesses, Frank Press, Bob Frosch, and myself. In the actual testimony, I essentially supported the administration's policy, praised the separation of the military and civilian space programs, and reiterated the importance of the shuttle for the conduct of certain missions related to the national security. The trouble arose in dealing with the written questions that Chairman Fuqua asked Frosch and me to answer. One of these went as follows: "One tenet of the Administration's space policy states: It is neither feasible nor necessary at this time to commit the United States to a high-challenge space engineering initiative comparable to Apollo. (a) Do you agree with this statement? (b) What do you consider to be appropriate long-term goals for NASA to undertake after the completion of the Space Shuttle?"

There it was. In order to answer this question, I would be forced to disagree with the administration's space policy, and I did so as carefully and in as limited a way as I could (see chapter 7). As it turned out, Frosch was also not very happy with the space policy and went some way to distance himself from it as well. Here is what he said in reply:

> Yes, I do agree at this time we do not need an Apollo-like initiative. My personal view is that we should be looking toward a significant national commitment to a major engineering initiative within the next three years. Candidates would include full exploitation of the STS (that is, the Space Transportation System as the Shuttle and all its ancillary equipment is sometimes called), intensive unmanned *in-situ* planetary exploration, space power systems, permanent space occupancy, perhaps extending to geosynchronous orbit and applications data systems.

Thus, Frosch said that the time for which the current space policy was valid would be coming to an end soon and that we would then have to think about a large new commitment. I took essentially the same

position except that I was somewhat more critical of the negative statement in our policy and that I was very clear on the matter of what
should be done next. Here is what I said in reply to the same question:

> I agree with the statement if it is interpreted to mean a project
> other than the Space Shuttle. We must push the Space Shuttle
> through to a successful conclusion. I believe that the choice of
> language in this statement may have been unfortunate since a
> negative statement of this kind should never be a part of a procla-
> mation of goals. The next major commitment after the Space
> Shuttle is completed should be the development of a permanently
> orbiting manned space station.

I did not know it at the time but what I said in reply to Fuqua's
question would occupy the largest fraction of my working life for the
four years to come.

On the same day of the testimony on space policy—July 24, 1980
—Dr. Albert D. Wheelon came by to see me. Wheelon was (and
still is) a senior vice president of Hughes Aircraft Company and
heads the company's communications enterprise. He is one of the
most thoughtful and successful leaders of the American aerospace industry and has made great contributions to all three areas of the aerospace
business—commercial, scientific, and military. I have known Wheelon
since we were both graduate students together at MIT in the early 1950s
and have worked with him over the years on a number of things, the
best being the Pioneer Venus Project that was managed for NASA by
Ames and for which Wheelon's division of Hughes Aircraft Company
was the prime contractor. I had (and have) complete confidence in his
discretion and judgment and that is why I felt completely free to talk
with him even about delicate matters. Wheelon came to see me because
he was worried about the shuttle program. At the time, the First
Manned Orbital Flight was scheduled for December 1980, and he
asked me to think about what might happen to the shuttle program if
Governor Reagan won the presidential election scheduled for November.
What Wheelon feared was that a first flight during the interregnum
between two administrations would, in all probability, be canceled
because the incoming president would want to be apprised of the risks
and then make his own judgment as to whether or not to proceed.
Thus, there would be further delays in a program that had suffered

more than its share already. Wheelon was well connected with some of the senior people in Governor Reagan's campaign, and he wanted to know whether I would be willing to provide some briefings for them on the shuttle program and its current status. He felt that if they were properly informed, then they could make better judgments of the risks involved. I told Wheelon that I thought he had a good idea, but that in view of the position that I held in the Carter administration, I would have to think carefully about his proposition.

I began to have doubts about President Carter's ability to win a second term in the summer of 1979. What triggered this feeling was the president's speech of July 1979 in which he said that the American people were afflicted with a "malaise." I thought that this was a particularly unfortunate turn of phrase because there is no way that you can lead people by telling them that they are sick. The events following the speech in the remaining months of 1979 and in the first half of 1980 confirmed my suspicions and Wheelon's visit forced me to make a choice. I worried over the matter for over a month, but, by the end of August, it was completely obvious that Governor Reagan would win the election, and therefore the contingency that Wheelon was worried about might very well happen. So, on August 28 I called Wheelon and asked him to go ahead and arrange the meeting. I did not tell Harold Brown that I was going to see some of Governor Reagan's people because I feared being told that I could not talk to them. I felt badly about this, but I did not know what else to do in view of the circumstances. (Perhaps I made a mistake because Brown probably would have told me to go ahead, but I was afraid to take that risk.) I salved my conscience by writing out a letter of resignation that I resolved to send to Brown in the event that President Carter won reelection. As things turned out, I could have saved myself the trouble since I would be relieved of my post in due course without having to resign!

Wheelon arranged the meeting, and it was scheduled for September 11, 1980. I had requested that he try to get Mr. Richard V. Allen to come to the meeting because I felt that Allen was the odds-on choice to become the national security advisor in the event of a Reagan victory. Since the national security argument had been successful in the Carter administration in keeping the shuttle program on track, I thought that the same might hold for a Reagan administration should there be one. Allen was thus the key man. Wheelon succeeded, and a dinner meeting was arranged at the Metropolitan Club in Washington, which would be attended by Wheelon, Allen, Mr. Clay T. (Tom) Whitehead, who had served in the White House during the Nixon administration, and myself.

By a really amusing coincidence, I had actually met Allen for the first time earlier that day. It turned out that Allen and I had both been invited to appear on the "McNeil-Lehrer Report"—the PBS news show —that same day. We were to debate the issue of the "stealth" bomber that was then part of the campaign debate. Allen was to represent Governor Reagan's position, and I would be defending the Carter administration. The debate went off well—at least both of us thought so—and we left the studio. When Allen, who is a man with a piquant sense of humor, saw me again at dinner, he was hugely amused. Where else, he said, could the antagonists of the afternoon become conspirators of the evening? The circumstances of our earlier meeting broke the ice, and Allen and Whitehead listened to me describe the status of the shuttle program in detail for almost four hours. At the end of the dinner party, Allen promised to help out if he could. As things turned out, he was true to his word and proved to be important in getting things on the right track for the space program during the first year of the Reagan administration.

X The Election of President Reagan

On November 4, 1980, former California Governor Ronald Reagan was elected to become the fortieth president of the United States. I cannot say that I was surprised. As long ago as the summer of 1979, I was very concerned about President Carter's ability to win a second term. On July 8, 1979, I put the following entry in my daily diary:

> This is the first day of our third year here in Washington. It has now been long enough to form some firm opinions about the Administration and what it is trying to accomplish. There is no doubt that the Administration is "honest" in the sense that Nixon's was not. Mr. Carter and Co. do not deliberately try to fool the people. In fact, they are almost painfully honest and this sometimes creates an impression of naivete and simple-mindedness that really hurts. To some extent, the Administration has proved that honesty is a necessary but not sufficient condition for good government—effective government I should say. The real trouble is that the Administration has not fired people's imagination. There is somehow a lack of leadership and excitement. Most important of all, the President has not created a credible vision of the future that people can believe in. This is the most serious failing and it may cost the President the next election.

Shortly after I wrote these words, President Carter made his "malaise" speech, and from then on I did not think that he would win a second term. Nevertheless, the actual realization that there would be a change of administration caused me great concern. There was so much that was left hanging. It is one thing to anticipate something, but quite another to actually see it happen. I was worried about two things. One was that the shuttle had not yet flown in space, and I feared that the new administration might delay the first flight beyond the then-projected date early in 1981. There was, after all, a substantial risk involved, and I could easily understand that they might want to conduct a detailed review before flying. Any further delays in the first flight would simply add weight to the arguments of those people who felt that the shuttle was not capable of meeting national security requirements. The other had to do with the Air Force and the completion of the west coast launch site at the Vandenberg Air Force Base. After the many arguments over this issue within the Air Force, I was afraid that there would at least be long delays or, at worst, possibly a cancellation of the program. What I wanted to do very badly was to stay on in Washington and see these projects through to the point at which I thought that they would be completed.

As things turned out, my initial worries about the schedule for the first shuttle flight were unfounded. Perhaps the most important immediate step taken by the new administration was to keep the top people in NASA in place until the first flight so that there would be continuity in the leadership until that critical hurdle was passed. Bob Frosch had announced that he would resign as administrator of NASA even before the election (October 6, 1980) and that he would be leaving in January 1981, no matter who became president. Alan Lovelace, the deputy administrator, was appointed acting administrator and was also given the title general manager to make sure that it was understood that the acting administrator was not just an interim appointee, but that he was someone who had the full authority to act under any circumstances until the first flight of the shuttle. The other senior leaders of the space shuttle development program, John F. Yardley, the associate administrator for space flight, his deputy, L. Michael Weeks, and Dr. Stanley Weiss, the associate administrator for space operations, were all kept in their respective posts.

Within the Air Force, things were not as straightforward. I would remain at my post as secretary of the Air Force during the period of transition between the old and the new administrations—that is, between election day (November 4, 1980) and inauguration day (January 21, 1981). However, it soon became clear that I would only have nominal influence over the course of events and that all I could do was to watch and fret when things went wrong. There was, of course, no chance at all that I would be retained as secretary of the Air Force by the new administration. I felt, however, that there might be a chance of being selected to one of the senior positions at NASA—either as administrator or as deputy administrator. If that could be arranged, then I would be in a position to help push the shuttle program through to completion and perhaps to help keep the construction of the west coast launch site on schedule as well.

The notion that I should rejoin NASA in one or the other of the two top positions was not new. As early as April 18, 1980, Bill Perry, after a meeting at which we discussed some of the problems we were having at the time with the shuttle, suggested that I should perhaps think in terms of moving over to head NASA in a second Carter administration should there be one. About two months before the election, I had agreed to do another assessment for the then deputy secretary of defense, W. Graham Claytor, about the status of the shuttle program and on the ability of the shuttle to meet the requirements of the national security related space program. In doing this job, I had asked for advice from a number of former NASA officials including former Administrator James C. Fletcher and Mr. James C. Elms, former deputy director of the Johnson Space Center and former director of NASA's Electronics Research Center in Cambridge, Massachusetts. I met with Fletcher and Elms on October 30, 1980, to go over the substance of the report I was preparing for Claytor. At the end of the meeting, we discussed the possibility of my moving to NASA in some capacity in the event of Mr. Reagan's election to the presidency. I told them that I would be most willing to serve in whatever capacity seemed reasonable. I also said that I did not think there was much chance of securing a political appointment at NASA because of my service with the Carter administration. We left things there to await the outcome of the election five days later.

Shortly after the election, I reactivated an application that I had submitted for a civil service scientific position at the Ames Research Center. I had always intended to return to California after my service in Washington—in fact, this is why I had prepared the application to Ames in the first place—and if my efforts to secure a high-level NASA post in the new administration were unsuccessful, then a return to Ames would be a logical move. I had held administrative posts for twenty years (starting with P-Division at the Livermore Laboratory in 1960), and I felt that perhaps the time had come to do something else. There was a complication in that by the end of 1980 my wife was about half way through her studies for a Doctorate in Education at George Washington University. Even though she too—as a native Californian —was anxious to return to California, she also wanted to finish her degree. Accordingly, I also made preparations to secure an appropriate civil service position in the Washington area that I could take up once I left the Air Force. I met with the director of the NASA Goddard Space Flight Center (Mr. A. Thomas Young) and the director of research of the U.S. Naval Research Laboratory (Dr. Alan Berman), and they both agreed that I could spend some time at their respective institutions. As things turned out, I would spend most of my time between leaving the Air Force (February 1981) and rejoining NASA (May 1981) in an office in the Space Science Building at the U.S. Naval Research Laboratory that was made available to me by Dr. Berman.

On November 14, 1980, I flew to California to address a meeting of the Air Force Association in Sacramento scheduled for November 15, 1980. I had arranged with Mr. James C. Elms to meet at Ames in the afternoon of November 14, and as it turned out, we also spent much of the morning on Saturday, November 15, together before I flew to Sacramento to make my speech. These conversations were critical to what subsequently happened, and this is why I want to record them in some detail. I have already introduced Jim Elms in these pages, but I should say more about him because he is a unique personality. Elms started his career in NASA as Robert Gilruth's deputy director when the then Manned Spacecraft Center (now Johnson Space Center) was being organized in Houston. Subsequently, he moved to Washington to work with George Mueller in the overall management of NASA's manned space flight program and, in 1967, he became the second director of the NASA Electronics Research Center located in Cambridge, Massachusetts. I first met Elms in early 1969 shortly after I became the director of the NASA Ames Research Center. (In fact, our first meeting was at the session of George Mueller's management council meeting during which I first learned about the space shuttle and the space station. See chapter 5). I took an immediate liking to him, and we have been friends ever since.

The NASA Electronics Research Center was established to create a competence in electronic component technology. During the early 1960s, a number of spacecraft were lost due to component failures, and it was felt that the government (i.e., NASA) should have a laboratory at which problems of the kind that were being encountered by the spacecraft could be solved. It turned out that by the late 1960s, the electronics industry had turned to and solved the problems that had bedeviled us earlier in the decade. Thus, by 1969, when the financial climate for NASA was beginning to close in, the NASA management began to look for institutions that could be closed in order to reduce manpower and the costs of operation. Early in 1970 a judgment was made that the NASA Electronics Research Center could indeed be closed because the work that was being planned for the center was no longer essential to the success of NASA's missions. It was during this critical time that Elms, as the director of the center, displayed his talent and resourcefulness as a manager and a leader.

I have already mentioned that when I agreed to join NASA early in 1969, the individual who persuaded me to take this step was Mr. James M. Beggs, who was then serving as NASA's associate administrator for advanced research and technology. Beggs had come to NASA from the Westinghouse Corporation a year earlier and was then in charge of managing the four NASA research centers—Langley, Ames, Lewis, and the Electronics Research Center. All of the center directors, including Jim Elms, reported to him. Although Beggs was the one who persuaded me to join NASA in 1969, I would not have the chance to work for him. As things turned out, he was asked to become under secretary of transportation by President Nixon early in 1969, so he left NASA before I actually joined the agency. Elms, however, had worked for Beggs for a year, and so he was well acquainted with him. When the NASA management made the decision to close the Electronics Research Center in 1970, Elms decided to strike out on his own. He saw that there were a number of capabilities that had been established at the Electronics Research Center that could be useful to the Department of Transportation, especially to the Federal Aviation Administration (FAA). Elms prepared a proposal to have the Electronics Research Center transfered from NASA to the Department of Transportation and submitted it to Beggs. Beggs bought the idea and developed a plan for the transfer. There is no doubt that Beggs is a very astute and accomplished political operator, and in due course he managed to work out conditions for the transfer of the Electronics Research Center from NASA to the Department of Transportation, which was successfully accomplished on July 1, 1970. Elms remained on as the director of the newly renamed Transportation Systems Center and worked for Beggs in that capacity for several years.

I was, of course, familiar with this story, and Elms argued that it would be important to have someone with the political and leadership skills of Beggs to head NASA during the next few years. We agreed that the next steps would have to be to make the shuttle operational and to get on with the development of the permanently manned orbiting space station that we had been talking about since the late 1960s. Elms also felt that Beggs would be just the man to develop the strategy to push this plan through the new administration and the Congress. Furthermore, Elms told me that he thought I would be the ideal deputy for Beggs. Beggs would provide the overall direction and also the proper political connections since he was well placed in Republican politics. I would bring some technical expertise as well as recent experience with NASA-Air Force relations that Elms thought might be valuable. Elms felt that with Beggs as NASA administrator and me as the deputy administrator, it would be possible to push through the things we had in mind. I agreed with him and told him that I would do whatever I could do to help bring about this outcome.

After these meetings on November 14 and 15, 1980, Elms went to former Governor John Volpe of Massachusetts who had served as secretary of transportation in the Nixon administration and was, in fact, Beggs' immediate superior during the time of the transfer of the Electronics Research Center from NASA to the Department of Transportation. Elms apparently persuaded Governor Volpe that the combination of Beggs and me to head NASA would be a good one. I do not know how much influence Governor Volpe actually exerted on this matter because I was very clearly on the "outside" during all of these negotiations as a member of the outgoing administration. It is also true that many other friends of mine exerted themselves on my behalf to secure a post for me in NASA. Among these were Jim Fletcher, Edward Teller; generals Bernard Schriever, Duward (Pete) Crow, and Tom Stafford; senators Barry Goldwater, John Glenn, and John Tower; Robert Gilruth, Albert D. Wheelon (Hughes), Professor Wilson K. Talley (University of California/Davis), Robert Anderson (Rockwell), Tom Jones (Northrop),

Donald D. Smith, and a number of others. Perhaps the most important of these was former NASA deputy administrator and later president of Rensselaer Polytechnic Institute, Dr. George M. Low. On November 22, 1980, I was in my office cleaning up some Saturday paperwork when Low called and informed me that he would be heading up the new administration's "transition team" for NASA. The function of these "transition teams" was to help each of the major federal agencies negotiate the change from one administration to another. Specifically, Low wanted to know whether I would accept the job as NASA administrator. I told Low that I would be most pleased to do that, but given the political circumstances, I did not think that it was a good idea. I then told Low about the plan that Elms and I had discussed a week earlier in California, and Low agreed that Elms's idea might be more congenial to the new administration and that he would think about recommending it to them. I was heartened by the call from Low. He was a highly respected figure in the space business, and although he was a Democrat, there was at least a chance that the new administration would listen carefully to what he had to say.

The next few months would be very confusing and frustrating. For example, on December 11, 1980, Mr. Jay Morris from the White House Personnel Office called to inform me that I was a candidate to become NASA administrator. A few minutes later, Jim Elms called me to tell me that Frank Borman, the former *Apollo 8* astronaut and currently the president of Eastern Airlines, had been offered the NASA administrator's post but had turned it down. Elms also said that Jim Beggs had been offered the NASA administrator's job as well and that he would probably accept. I did not hear from Morris again until January 17, 1981, when he told me once again that I was still a candidate for the job. In the meantime, there were numerous other rumors about who would head NASA and a large number of names appeared — including my own. On January 29, 1981, there was an item in the "Executive Notes" section of the *Washington Post* that both senators Barry Goldwater and Harrison (Jack) Schmitt were supporting me for the NASA administrator's job. Jim Beggs was also mentioned in the same column. I would not find out what was really going to happen for two months.

The last day of the Carter administration came on Tuesday, January 20, 1981. We watched the inauguration ceremony of the new president,

Ronald Reagan, on the television set in my office. My secretary, Mrs. Jo Watson; my military assistant, Colonel (now Major General) Craven C. Rogers; and my old friend, Tom Stafford, who had by now retired from the Air Force and was pursuing business interests in Oklahoma, were with me. The inauguration was really very well done. Mr. Reagan broke precedent by using the "back" side—that is, the west side—of the Capitol building from which to make the inauguration speech. This side of the building overlooks the mall and really is a much better place for the inauguration than the "front" side. Mr. Reagan made an excellent speech. He told the American people that we were great and strong and that we should once again begin to act that way. There is no doubt that the speech was very inspiring and that it created an entirely different political atmosphere than the one prevalent for the last two years of the Carter administration. I hoped fervently that Mr. Reagan's optimism would be contagious and that the nation would recover its self-confidence.

Three days after the inauguration of President Reagan, the name of my successor as secretary of the Air Force was announced. The post would go to Mr. Verne Orr, who had served with distinction as director of finance in California under Governor Reagan. I was pleased with the choice because Orr had an excellent reputation as a public servant, and he was also close to the new president, which was something that might be of great value to the Air Force. I met with Orr for the first time on January 24 for an intensive series of briefings on the problems we were facing in the Air Force. Orr proved to be extremely intelligent and quick on the uptake and took charge quickly. He has become a very strong secretary, and he and I have become good friends over the years we have worked together. Orr was confirmed in his new post by the Senate and assumed office on February 9, 1981. Mr. E. C. (Pete) Aldridge, Jr., was to become under secretary of the Air Force.

I moved out of my office and for a couple of weeks was permitted to use an office down the hall. Then I moved over to the U.S. Naval Research Laboratory to the office that my friend Alan Berman (the director of research at the Naval Research Laboratory) had provided. I spent most of my time at the Naval Research Laboratory working on a paper titled "Technology and the Strategic Balance," which I would use as a basis for the Charles H. Davis Lecture that I had been asked to deliver at the Naval War College later on in the spring of 1981. The paper was actually based on a longer paper that I had prepared together with Air Force Colonel John Endicott in which we made an attempt to redefine the strategic position of the United States in the world in terms of some of the technologies that are now emerging. What is perhaps most interesting about the "Technology and the Strategic Balance" paper (which was published later on in 1981) is that I explicitly mention the possible development of defenses against ICBMs and SLBMs that would depend on advanced technology (highly accurate guidance systems and airplane mounted lasers) then under development rather than the ABM systems of the late 1960s that required nuclear warheads to kill the target. This was to become a controversial issue later on in the Reagan administration.

The rumors and maneuverings continued when I moved over to the Naval Research Laboratory, but I slowly came to the conclusion that nothing would come of the effort to secure a post for me at NASA. Accordingly, I began to make other plans. I reached the decision to stay in Washington because of my wife's school situation, and I applied for a permanent position at the Naval Research Laboratory to replace Dr. Herbert Friedman as superintendent of the laboratory's Space Science Division. (Friedman had just retired after a most distinguished career in that post.) Now, I could do nothing but wait and see what would happen.

There is one event that occurred during my stay at the Naval Research Laboratory that deserves special mention. I had a telephone call from the office of Congressman Newt Gingrich (R., Georgia) asking me to meet with the Congressman. A meeting was scheduled for February 26, 1981, and at the appointed time, I appeared at Gingrich's office. Gingrich is a former history professor with a very sharp and comprehensive intelligence. His first question to me was both simple and sweeping. He said: "You have now served in the Pentagon for four years and I want to know what you have learned during that time!" I have to confess that I was taken aback, and I had to mumble and make excuses while I was collecting my thoughts. Upon reflection, the question was an excellent one, and it speaks well for Gingrich that he asked it. (I am surprised, in fact, that he was the only one of the 535 members of Congress who bothered to ask such questions.) I tried to answer as best I could, and our first meeting has turned into what has become a strong and productive relationship. Gingrich is a strong supporter of the space program because he believes that it is an essential element of the vision for the future that this nation must have. This is a well thought-out position with Gingrich, and he has expressed it in detail in his recently published book, *The Window of Opportunity*.

My first chance to meet some of the members of the new administration socially came on March 1, 1981. Throughout my term of service as secretary of the Air Force, I had made a habit of having lunch periodically with the former secretaries who were resident in the Washington area, John L. McLucas and Thomas C. Reed. Tom Reed had served as secretary during the Ford administration and had also spent some time in other high Pentagon posts. Prior to his Washington service, Reed had served on Mr. Reagan's staff in Sacramento during the latter's term of office as governor. I had known Reed slightly from Livermore days when he had spent some time at the laboratory, and during our more recent lunch meetings, it quickly became apparent that we saw eye-to-eye on many issues that were facing the Department of Defense at the time.

March 1 was Tom Reed's birthday, and a party was arranged for him at Wexford, which is a large country estate near Middleburg, Virginia. At the time, Wexford was owned by then Texas Governor William P. Clements who is one of Reed's friends, but Wexford was also the house that John F. Kennedy bought shortly before he became president to use as a "retreat" near Washington. Pictures and photographs of Kennedy and his family were still hanging in the main hall of the house as he had left them, and it was a strange experience to stand in this house almost twenty years after the fateful event in Dallas that had destroyed the owner.

The two most important guests at this party were the new deputy secretary of state, William P. Clark, and his wife, and Michael K. Deaver, and his wife. Deaver would become the deputy chief of the White House staff. Both Clark and Deaver would become very influential members of the new administration and supporters of the space program. Clark would play a particularly important and positive part in developing the new administration's space program, but this was still more than a year in the future. At the time of Reed's birthday party, I was, of course, still completely uncertain about my own future, but I nurtured the hope (a false one, probably!) that attending events such as this one might be helpful in my effort to secure a post at NASA.

During the period I spent at the Naval Research Laboratory, I continued to have many conversations with my friends who were working to secure a post for me in NASA. Jim Elms continued to be the most active of these, but there were many others as well. Obviously, I do not know how the debate about my joining NASA went within George Low's transition team and in the White House political personnel office. As a member of the previous administration, I could hardly expect to be taken into their confidence. Whatever took place, the idea that Jim Elms and I had discussed the previous November apparently took root. On March 13, 1981, I received a telephone call from Mr. Pendleton James, who was at that time the chief of the political personnel office in the White House, and he informed me that the president would like to nominate me to serve as deputy administrator of NASA. He also told me that Jim Beggs had been asked to serve as administrator of NASA. Needless to say, I was elated. I would now have the opportunity to try and finish what had been started in the Carter years, and this was most gratifying. I felt a deep sense of obligation to those people who had helped me secure my new post, and I was determined to make the time I would be spending at NASA as productive as possible.

The first order of business was to meet with Jim Beggs to decide what we should do once we assumed our respective offices. On March 21, 1981, I flew to St. Louis where Beggs was living at the time. (When he was selected to become NASA administrator in March 1981, Beggs was serving as the executive vice president of the General Dynamics Corporation with headquarters in St. Louis.) It did not take us long to make up our minds on just exactly what our priorities would be during our term of office. Fortunately, Beggs had served in NASA in 1968, which was the year that many of the plans for the space station and the space shuttle had come to fruition. He was, therefore, thoroughly familiar with what had been done and with the thinking that had governed the planning. We determined that we would do the following things:

1. We would work hard to turn the space shuttle into a fully operational space transportation system.

2. We would try to persuade the new administration to adopt the construction of a permanently manned orbiting space station as the next major goal in space.

3. We would do something about the upper stage problem that was then (and, I am afraid, still is now) plaguing us.

4. We would make some major organizational and personnel changes.

For the longer term future, the most important of these objectives was, of course, the space station. Under the space policy of the Carter administration, the initiation of the space station program would not have been possible because of the provision in the policy that it was not deemed desirable to start large new engineering initiatives (see the precise quote on page 80, chapter 8). The Carter policy used the words "Apollo-like" to define the scale of the programs that were included in this proscription. Even though the space station we had in mind was not nearly as large as the Apollo program, we were certain that our prospective opponents would still characterize it that way. We, therefore, determined that we would take the first opportunity we had to tell people what we had in mind with respect to the space station. We wanted to establish this goal as early as possible so as to smoke out the potential opposition. We also wanted to make certain that there would be no blanket proscription against ambitious goals in the space policy of the new administration. The best way to do this was to state our own goals and objectives as clearly as possible and to see whether anyone objected.

The first opportunity to do this came at our confirmation hearing before the Senate Subcommittee on Science, Technology and Space of the Committee on Commerce, Science and Transportation. Senator Harrison (Jack) Schmitt (R., New Mexico) was now the chairman of the subcommittee since the Republicans had gained control of the Senate in the 1980 elections. We felt certain that at some point during the hearing we would be asked what we thought should be done next. Sure enough, Schmitt asked the right question, and both Beggs and I responded by saying that the construction of the space station was the next logical step. We had sown the seed and now it was squarely up to us to make it grow.

Three weeks after I received the call from Pendleton James about my new appointment, the space shuttle *Columbia* was finally ready for her first flight. I was in a somewhat awkward position early in April 1981. Although I had been informed that I would become NASA's deputy administrator, there had been no formal public announcement. At the time of the scheduled launch, therefore, I was in a kind of political limbo that made it difficult for me to participate. I finally called Dr. Christopher C. Kraft, then director of the Johnson Space Center in

Houston, and asked him whether I could watch the launch from the Mission Control Center. Kraft, who knew of my prospective appointment, readily agreed to my request. I had a very special reason for making this request. I wanted to watch *Columbia*'s first flight from the same spot where I saw Neil Armstrong's landing on the moon nearly twelve years before. Kraft also very kindly agreed to let me sit in the control room itself rather than the viewing room, which is divided from the control room by a glass partition. I would, therefore, have the great pleasure of watching *Columbia*'s historic flight together with my old friends Aaron Cohen and Milton Silveira since I was slated to sit together with them at the console assigned to the Shuttle Orbiter Project Office.

I went to Houston on April 9, 1981, to participate in the countdown for the launch then scheduled for April 10. Unfortunately, the backup flight computer failed, and so the scheduled launch had to be scrubbed. We spent April 11 diagnosing the problem and developing an appropriate solution without too much trouble. The launch was rescheduled for Sunday morning, April 12, 1981, at about 7:00 A.M., EST. On this attempt, things went perfectly. It was an awesome sight to watch the great white bird lift off from its pad, roll over to the correct azimuth, and then disappear in the distance. It had been a long time in coming, but now we were on our way. Twenty minutes later we knew from the data display on our console that everything had worked perfectly and that *Columbia* was now in her planned circular orbit at an altitude of 130 nautical miles. The crew, mission commander John Young, and pilot, Robert Crippen, were elated. We could hear that from the tone of voice they used in making their reports. Even John Young was excited. As a quintessential possessor of the "right stuff," Young is not ordinarily a voluble person, yet even he had a hard time keeping the triumphant edge in his voice under control.

A little more than an hour after the successful lift-off, we prepared to open the payload bay doors. This was the first scheduled operation on orbit, and it would, of course, give us our first good look outside through the television monitor cameras that were located in the payload bay. We watched intently as the doors slowly opened—first one, and then the other. It was definitely a beautiful sight to watch the earth moving underneath the *Columbia* as she traveled in orbit. However, when the cameras panned around to look back toward the aft end of

Columbia's payload bay, we were in for a shocking surprise. When we looked more closely at the view of the two pods containing the orbital maneuvering engines that were mounted on either side of *Columbia*'s tail, there was clear evidence that several of the tiles of the Thermal Protection System mounted on the pods were missing. This was, indeed, a shocking discovery. After all of the problems that we had with the tiles and all of the lengthy and painful effort that had gone into solving them, we apparently had not succeeded. Furthermore, and this was the real question, if there were tiles missing on the pods, where we could at least see them, were there tiles missing elsewhere on *Columbia*? More important, were they missing from regions of the surface where such a failure could possibly be catastrophic?

We did not know the answer, nor was there any way we could find out. The only thing we could do was to perform the best calculations we knew how to make at the time and to try and convince ourselves that it was indeed safe to bring *Columbia* in for her scheduled landing at Edwards Air Force Base on April 14, 1981. We determined first that the missing tiles on the pods containing the orbital maneuvering engines would not present a problem. As it turned out, there was still enough insulation left on the surface of the pod that it could withstand the heating that we expected there during the entry sequence—at least this was true according to the best calculations that we could make. The more important question was whether there were any other tiles missing elsewhere on *Columbia*. All we could do was to speculate. We felt that the reason the tiles came off on the pods had somehow to do with the unusually large aerodynamic stresses that might be present there because of vortex formation during the ascent trajectory caused by the delta wing configuration of the vehicle. Such vortices would create unusually large forces on the tiles, large enough, perhaps, to make them come off. There were no forces that we knew of elsewhere on the vehicle that would be exerted as large as those due to the potential vortex formation during the ascent trajectory that I have discussed. Furthermore, the tiles that covered the regions of *Columbia* that were critical to the entry sequence were "densified" or strengthened to withstand possible aerodynamic shear forces as well as other stresses.

We spent almost all of April 13 in a concerted effort to come to grips with this problem. There was, of course, not much that we could really do. If there were indeed tiles missing from the portion of *Columbia* that would experience the highest heating rates, then we would only find out by making the attempt to bring her back to earth. So we made our best calculations and estimates, but, in the final analysis, we had to place our trust in the Lord and hope for the best.

Columbia landed successfully on the dry lake bed at the Edwards Air Force Base a little bit after noon, local time, on April 14, 1981. It was a triumphant moment. We had kept the promises that we had made back in 1972 when President Nixon had originally approved the space shuttle program. We had built a reusable space ship; we had done the job for a little more money than we were originally given—eight billion dollars rather than six—and, we were over two years late in executing the first flight. While we did not fulfill all of the promises, we came close.

The ceremony after *Columbia's* first safe return from earth orbit was a really triumphant moment. The event was broadcast in its entirety over the NASA Net so we could see it on the television monitors in the Mission Control Center. I still vividly remember Al Lovelace standing on the podium, lifting his arms in a broad gesture, and announcing, "We're number one again!" Indeed, we were. After a hiatus in the manned space flight program of seven years, we were back in business. I felt genuinely pleased that it fell to Al Lovelace to make the announcement of our success. This was altogether fitting and proper since he had done so much to bring that success about in the first place. After the landing ceremony was finished, Max Faget came over to our console and said to us with the broadest of grins on his face: "Let's do it again!" That was, indeed, what was new. With *Columbia*, we could, for the first time, do it again, and we could repeat the feat over and over. Faget had caught the whole point of the effort in one simple statement.

XI

The Effort Starts in Earnest

As luck would have it, the public announcement that I would be nominated to become deputy administrator of NASA was made on April 24, 1981, on which day I happened to be making a visit to my old laboratory, the NASA Ames Research Center. I thought that this was entirely fitting and proper and that it was perhaps even a good omen for a successful tenure in office. As I have already said, our confirmation hearings were held on June 17, 1981 (my fifty-second birthday), and Beggs was sworn into office by Vice President Bush on July 10, 1981. I was sworn in by Beggs later on that same day. We were now officially on our way. One of the first things I was asked to do was to testify on *Columbia*'s first flight before the Congressional committees that had the responsibility to deal with the national security related space programs. Having made many promises about the shuttle during the past four years in front of these same groups, it was a great pleasure for me to stand before them now and present the actual evidence that the shuttle was indeed a technical success. Several senior members of these committees, including Senator Barry Goldwater and Representative Edward P. Boland had been very strong and consistent supporters of the shuttle during the difficult years when we were trying to sustain the program in the face of technical and financial problems. It was a special privilege for me to stand before these distinguished members of the Congress and to tell them that their faith in us was justified.

I had moved to a temporary office at the NASA headquarters building on Maryland Avenue in May to start things going. In May and June, there were many talks with numerous people, but the most important of these was with Philip E. Culbertson, who was an old friend from my

Ames days. Culbertson is one of the very best people at NASA, with many years of experience in manned space flight. During the Carter years, Culbertson served as a special assistant to Administrator Robert Frosch and was now interested in playing a major role in the space station project. Culbertson and I felt that the best way to do this would be to persuade Jim Beggs to appoint Culbertson to the post of associate deputy administrator, which is the number three position in the NASA hierarchy. Because of the importance of the space station program, we felt that the person who would be in overall charge of organizing the effort in NASA should occupy this post. Culbertson and I also agreed that now was the time to make the major push to get the space station adopted as the next major initiative in the American space effort. Beggs, who started work full-time in Washington in mid-June quickly agreed, and Culbertson was appointed associate deputy administrator and given overall responsibility for organizing the preliminary engineering studies to look at the feasibility of the space station ideas that were then being developed by the Johnson Space Center and the Marshall Space Flight Center. Culbertson would turn out to be a key figure in the effort to initiate the space station program.

Another old friend from my days at Ames who agreed to join me in Washington was Dr. Milton A. Silveira. Silveira was serving as the deputy director of the Shuttle Orbiter Project Office at the Johnson Space Center when I was appointed deputy administrator. I felt strongly that we needed to have someone with firsthand experience in the technical areas important to the shuttle program close to Beggs and me. Accordingly, I asked Silveira to come to Washington to serve as a special assistant to the deputy administrator. This turned out to be an extremely important move because of the technical problems we would be facing in the early flights of the shuttle with things like the auxiliary power units, the fuel cells, and a number of other items.

During July and August, there were a number of meetings of the senior staff at NASA headquarters during which Beggs stressed the importance of developing a good long-range plan for the next four years that would form the basis for the programmatic decisions that we would have to make. In response, Silveira and I prepared a short paper titled "Notes on Long Range Planning," which we submitted to Beggs on August 18, 1981. In this paper, we outlined the basic priorities that we thought were important. First, we felt that we would have to pay much attention to the NASA institution. Ever since the peak of the Apollo effort in the late 1960s and early 1970s, there had been a steady decline in NASA's Civil Service manpower from 36,000 employees in 1967 to 22,000 in 1981. We were concerned about this erosion and wanted to stop it. We also put the maintenance and the development of NASA's aeronautical facilities at a very high level of priority in order to maintain NASA's ability to provide strong support for the nation's civil and military aviation programs. We thought that this was particularly important in view of the heavy commitments that the new administration was likely to make in new aircraft developments. (The B-1B bomber and stealth technology aircraft.)

Next in priority, we called for a strong push to make the space shuttle operational. We had to fulfill the promises that NASA made when the shuttle program was started in 1972. Although the first flight of *Columbia* in April demonstrated that the shuttle was technically a success, we still had to prove that it was operationally successful. This meant that we had to show that we could meet the flight schedules and payload requirements demanded by the users of the launch services provided by the shuttle. We also had to learn to control the operational costs. In this section of the paper, we made the controversial recommendation that shuttle operations should eventually be moved from the Johnson Space Center to the Kennedy Space Center. Third, we made the unequivocal recommendation that NASA should make a strong effort to persuade the new administration to initiate a program to construct a permanently orbiting manned space station. Finally, we recommended the continuation of NASA's space science and applications programs with some minor changes.

The long-range planning paper was submitted to Beggs and was also circulated among the senior NASA management people, both in Washington and at the various NASA centers. A few weeks later, Beggs returned the copy of the paper I had sent him with a brief note clipped to it saying: "OK, let's go." We were on our way. (A copy of "Notes on Long Range Planning" is included in appendix 4, together with Beggs's note.) Subsequently, after thorough discussions with a great many people inside and outside NASA, a comprehensive set of goals and objectives was developed that was based essentially on the August 1981 paper. These were issued in final form by Administrator Beggs on August 24, 1983. The important point about these planning documents is that they formalized the decision Beggs and I had reached in March to make the space station the next major initiative in space. In so doing, we committed the NASA institution to work toward this objective early in the new administration.

One amusing incident occurred at this time that deserves special mention. Since "Notes on Long Range Planning" was fairly widely distributed, it was likely that it would leak to the press. Sure enough it did. I received a call from Mr. Craig Covault, who is a distinguished journalist on aerospace matters writing for *Aviation Week and Space Technology*, asking for an appointment to talk about the document. I met with Covault on November 10, 1981. Much of the conversation with Covault dealt with the recommendation that shuttle operations be moved from the Johnson Space Center to the Kennedy Space Center since this was sure to be a matter of controversy. I tried to talk a little bit about the space station as well. While Covault was personally very interested in the subject, the major topic mentioned in his story and those that were derived from it had to do with the Johnson-to-Kennedy move. I would have been much happier if the press had paid more attention to the space station.

Developing an operational space shuttle system was the first programmatic priority that we felt was most important in the summer of 1981. We believed that in order to commit the administration to the shuttle program and the things that would follow, we should get some kind of statement from the president that the new administration strongly supported the shuttle program. We calculated that this would be easier to do now that *Columbia* had made her maiden flight than it was two years ago when we had to secure the same commitment from President Carter. It was important to get such a commitment not only for the purpose of continuing the shuttle program but also to continue the construction of the west coast shuttle launch site at the Vandenberg Air Force Base in California. The Vandenberg issue had once again come to the fore because of cost problems that were being encountered. On June 10, 1981, there was a meeting in the White House Situation Room to discuss the ways and means that would be employed to get such a commitment. Mr. Richard V. Allen, the president's national security advisor (and my dinner companion on September 11, 1980, see chapter 9), was in the chair. The other participants were the secretary of the Air Force, the Honorable Verne Orr; the president's science advisor, Dr. George A. Keyworth II; the deputy director of OMB, Mr. Edwin Harper; and the OMB's assistant director for national security, Dr. William Schneider. Also present was Colonel Michael Berta, USAF, who had replaced General Rosenberg as the chief space advisor on the National Security Council staff.

From the beginning of the meeting, it was completely apparent that Allen was very much in favor of making a strong commitment to the shuttle program. There was very little debate on this general issue, and the meeting broke up after agreeing to do the following things:

1. We would persuade the president to reaffirm the national commitment to continue the space shuttle program.

2. We would set up an appropriate decision-making process within the National Security Council staff to resolve day-to-day problems between NASA and the Department of Defense.

3. We would set up a launch vehicle study under the direction of Dr. Keyworth to look at the longer range strategy for how the shuttle and other U.S. launch vehicles would be employed. (I made this suggestion, but it was never implemented.)

Mike Berta was charged with following up the implementation of these decisions.

On July 7, 1981, there was another important meeting in the development of the new administration's position on the shuttle. This was a joint meeting of senior Air Force and NASA officials to review the progress of the construction program of the west coast shuttle launch site. Since the format of this meeting was cast in the style of an Air Force Program Acquisition Review (PAR), and since NASA as well as Air Force people were present, this meeting was dubbed the "Super PAR." Both Verne Orr and Jim Beggs attended this meeting, as well as most other senior Air Force and NASA officials. It was very obvious from the beginning of the Air Force briefings that there were serious technical and management problems that would eventually cause a delay in the projected completion date of the Vandenberg facility. My main concern as a result of this meeting was to minimize the length of the delay. On July 8 secretary Orr decided to delay the date of the initial operational capability of the shuttle launch site from October 1984 to October 1985. This was reasonable enough, given the circumstances that were described at the "Super PAR" and, at this writing, our plan is still to conduct the first shuttle launch from the west coast site late in 1985.

I had believed for a long time that there should be much closer relationships between NASA and the Air Force, especially on matters such as the construction of the shuttle launch site at Vandenberg. My concern was heightened by the "Super PAR" meeting. Consequently, I argued that it would be wise to have an experienced Air Force general manage the shuttle program during its initial phase of becoming an operational system. Fortunately, we had the ideal candidate in Major General (now Lieutenant General) James A. Abrahamson. Abrahamson is one of the most brilliant officers in the American military, and he had just completed a term managing the Air Force's F-16 fighter aircraft program. This program was, by any measure, the best run development program in the Air Force. Both Beggs and I knew Abrahamson well, and we had complete confidence in his ability to manage the shuttle program in such a way that it would become a credible operational system. We asked Abrahamson to do the job, and late in 1981 he became NASA's associate administrator for space flight, a post he would hold for the next two and a half years.

There was a meeting of the National Security Council on August 3, 1981, which Jim Beggs attended. He told us afterward that things went well and that we would be getting a strong statement from the president on the shuttle program. I subsequently met briefly with Allen a couple of times and also worked with Mike Berta to get the statement for the president in good shape. All of this effort finally came to fruition when the final statement was issued on November 13, 1981, as National Security Decision Directive-8 (NSDD-8) by President Reagan. It was an excellent statement, not only because it supported the shuttle program but also because it stated explicitly that there would be no changes in the content of the program without the explicit approval of the president, thus signaling clearly that the president had a strong personal interest in the space shuttle program. This was, indeed, much better than we had expected.

At this same time, we also began to have general conversations about the overall strategy that we would pursue to secure the approval of the administration for the space station program. I have already described the earlier conversations with Phil Culbertson, and these were now expanded to include other people as well. Former administrator James C. Fletcher was a major participant, so was Jim Elms, and finally Jim Beggs himself led the discussions. Beggs is nothing if not an extremely astute political manager. He recognized before any of the rest of us that, ultimately, there was only one person who would have to be convinced that the space station was a good idea, and that person was the president of the United States. Beggs therefore determined that our strategy for persuading the administration to pick up the space station as a new initiative would consist of two major elements:

1. We would focus our efforts of persuasion on the president and his immediate staff.

2. We would use the very favorable public reaction to the space shuttle missions to draw the president's attention to the possibilities of what could be done of political value with the space program.

I have to confess that, in the beginning, I was not quite as pessimistic as Beggs about securing support from other federal agencies and groups interested in space flight. I thought that it might be possible to garner support for the space station elsewhere in the government so that the direct approach to the president would not, in the end, be necessary. As things turned out, I was wrong and he was right. I spent much time trying to gain support for the space station program from the other agencies of the federal government and from the scientific community. I failed almost completely. Beggs had read the history of efforts such as this more accurately than I had—or at least he understood it more clearly. The Apollo program was undertaken by President Kennedy and the shuttle program was undertaken by President Nixon, both against the recommendation of their most senior advisors. I knew this, of course, but I kept hoping against hope that this time things might be different. I was wrong, and I shall describe the failure of my efforts in some of the forthcoming pages of this volume.

The first opportunity to interest the president directly in the shuttle program was not long in coming. The event in question here was the second flight of *Columbia* scheduled for November 1981. It would become our habit during the shuttle launches that Jim Beggs would go to the Kennedy Space Center and I would go to the Johnson Space Center for the launch and the flights. In this way, both of the important operations centers during the shuttle missions would be covered. The launch of STS-2 (the abbreviation then in use for the second shuttle flight) was scheduled for the morning of November 12, 1981. I was sitting at the Orbiter Project Office console, which was next to the Flight Operations Division console that was occupied by Gene Kranz (the division chief) and by Chris Kraft (then the director of the Johnson Space Center). It was about 6:00 A.M. local time—about three hours before the scheduled time for the launch—a quiet period, and Kraft was reading the local newspaper. Suddenly, he turned toward me, gave me the paper, and said: "Here, look at this." What he pointed to was a story announcing a visit to Houston by President Reagan and Vice President Bush. He suggested that I should get on the telephone and see if I could not get the people in Washington to persuade the White House people to schedule a visit by the president to the Mission Control Center during his stay in Houston. Kraft said that it would probably not be too hard to arrange things so that the president could talk to the crew using *Columbia*'s voice-link communication system. I called Beggs in Florida and told him about the suggestion that Kraft had made. Beggs agreed that we should try to get the president to visit the Mission Control Center and turned the problem of developing the plan to do this over to Jim Fanseen.

I should pause to say a word here about Mr. James F. Fanseen and his work during the years (1981–84) of the effort to persuade the president to make the commitment to the space station program. Fanseen is a member of a prominent Baltimore family and has a law practice in that city. He is also active in Republican politics and served as an "advance man" for President Reagan during the 1980 election campaign. Fanseen and Jim Beggs were friends of long standing, and when Beggs was selected to become administrator, he asked Fanseen to join NASA and become the liaison officer with the White House. This turned out to be an exceedingly important assignment because the strategy that Beggs was developing at the time to sell the space station program to the White House depended heavily on having as many access routes to the president as possible. Fanseen did a very effective job of keeping the communications channels open, and this would be his first opportunity to see how well he could operate.

Columbia's second flight started successfully a little after 9:00 A.M. (Houston time) on November 12, 1981. By this time, Fanseen was already hard at work on the arrangements. I spent most of the day following *Columbia*'s progress in orbit. At about 3:00 P.M. on November 12, we learned that there was a problem with one of the fuel cells (i.e., batteries) that supplied *Columbia*'s electrical power. Although the problem was not serious enough to cause a mission abort, we did decide to shorten the mission from the five days originally planned to three. I spent an hour on the telephone with Fanseen explaining the situation to him and then spent the rest of the day on the fuel cell problem. Shortly after midnight, Fanseen called again and informed me that the president would indeed be visiting the Mission Control Center at about 6:00 P.M. on November 13, 1981. He also informed me that Jim Beggs would be riding from Washington to Houston on Air Force 1. Finally, Fanseen told me that a White House "advance man," Mr. Steve Studdert, was on his way to Houston and would be arriving shortly. This meant that I would have to wait up for Studdert before I could get to bed. Studdert finally showed up shortly after 1:00 A.M. (November 13, 1981), and we spent an hour going over the Mission Control Center planning for the president's visit. It was well after 2:00 A.M. before both Studdert and I finally got to bed. Fanseen had done a neat piece of work. In a matter of a few hours, he had mobilized the White House staff, changed the president's schedule, and arranged for a visit by the president to the Mission Control Center. This visit would have far-reaching consequences as far as I was concerned since it marked the beginning of the president's personal interest in NASA and in the space program.

After spending most of the day working on the fuel cell problem, which had still not been resolved, we made preparations for the president's visit late in the afternoon. The president entered the Mission Control Center shortly after 6:00 P.M. accompanied by Jim Beggs, Michael Deaver, and several other senior White House staff members. Chris Kraft presented a short lecture to the president and his staff about the Mission Control Center and its functions. It was well done, and there is no doubt that the president was very clearly interested in what was going on. He asked a number of questions and then a communications circuit was patched together so that the president could talk with the crew, mission commander Colonel Joe H. Engle, USAF, and the pilot, Captain Richard H. Truly, U.S. Navy. President Reagan spent about four minutes in contact with the crew after which contact was lost as *Columbia* passed out of range of the Buckhorn ground station through which the communications circuit was patched. After the conversation with the crew, the president made the rounds in the control center and shook hands with most of the people sitting at the consoles. It was clear that he was in a good mood, that he was genuinely interested in what he saw (he spent about forty minutes at the Mission Control Center, rather than the twenty minutes originally scheduled), and he very much enjoyed himself.

I had met President Reagan once before in the early 1970s when he was the governor of California and I was the director of the Ames Research Center. Governor Reagan had asked all of the heads of the larger federal installations in California to meet with him in Sacramento. I went to this meeting prepared for a pro forma session and was quite surprised at what actually happened. The governor met with us (there were about fifteen people in the group) in his conference room and told us briefly that we represented a substantial payroll (that is, our employees really were the payroll) to California, and he wanted to know what we were up to. He asked each of us to talk for five minutes or so about our installation and went around the table. He listened, asked some good questions, and after we had each had our turn, adjourned the meeting. The surprising thing about the episode was that about two weeks later one of the governor's assistants called me to follow up with some more questions and also to offer help on a couple of requests I had made during my short remarks. I never expected either the governor's expression of interest much less the follow-up telephone call. The whole episode made a very positive impression on me, which was now reinforced by what I saw of him in the Mission Control Center. Perhaps the best way to describe how I felt about the impression President Reagan made on me is to repeat here what I wrote the next day in my daily diary (November 14, 1981):

> One more comment on the President's visit yesterday. It is very obvious that Mr. Reagan *enjoys* being President in contrast to his predecessor who really did not like the symbolic part of the job. I remember seeing Mr. Carter on similar occasions and he was brisk, businesslike, but, unfortunately, rather cold. Mr. Reagan, on the other hand, radiates warmth and really went out of his way to talk with as many people in the room as he could find. Perhaps this is the substance of the "star" quality I tried to define yesterday.

The one thing that was clearly apparent as a result of the president's visit to mission control was that he was genuinely interested in the space program. After November 13, 1981, I began to believe that Beggs's strategy of going directly to the president to persuade him that the space

President Reagan talks with orbiting astronauts. From left are spacecraft com-
municators Terry J. Hart (standing) and Daniel C. Brandenstein (seated),
Mark, James M. Beggs, and Christopher C. Kraft, Jr.

station should be built actually had a good chance of succeeding.

In the summer and the fall of 1981, there were many conversations
about what the space station we were proposing should actually look
like. There was by no means complete agreement within NASA on this
issue. Broadly speaking, there were two schools of thought about what
should be done, one represented by the people at the Johnson Space
Center and the other by those at the Marshall Space Flight Center. It is
worth discussing these viewpoints in some detail because they each
represented opposing, yet valid, views of what should be done.

For some years, the Johnson Space Center had been working on
plans for what they called a "space operations center," or the SOC. This
was a large space station with many capabilities, and it was something
that would clearly be very expensive. The central feature of the SOC was
that it would be permanently manned. In contrast, the approach advocated by the Marshall Space Flight Center was more evolutionary. The

people at Marshall wanted to build "space platforms" that would initially be unmanned but man-tended. That is, the platforms would be visited occasionally by the space shuttle to be "tended," but they would not be permanently manned. In due course, the space platform advocated by the people at Marshall would be expanded into the same permanently manned space station as the one advocated by the Johnson people. In spite of this, there were very genuine differences between the two approaches that had to be properly resolved. Even though Jim Beggs and I were thoroughly committed to a permanently manned space station, we were really not certain at the time which approach was best.

It might be worthwhile at this point to discuss in some detail what we actually had in mind at the time. We felt that it would be most advantageous for people to be present at the very beginning on a permanent basis in order to fulfill the three major functions we had postulated for the space station. The space station would be a laboratory in space, and it was clear to us that a really good and capable laboratory requires the presence of people. It is true that certain functions could probably be automated, but the fact is that for the laboratory to function, there is really no substitute for the presence of human judgment and imagination on the spot. The space station would be a maintenance and a repair base on orbit. Here again, we believed that only if people were actually present on a permanent basis would it be possible to develop this capability. Repairing and maintaining satellites that have failed or require maintenance takes human judgment and, once again, people on the spot would be essential in our view. Finally, the space station we had in mind would be a staging base for more ambitious missions that we would want to carry out in the future. Here again, the preparations for new manned lunar landings or possibly a manned mission to Mars would clearly benefit from the presence of people to do these jobs.

While we were fairly sure that the approach I have just outlined was correct, we were still not quite certain. We therefore felt that we needed to ask for some outside advice on this subject. We were particularly concerned about the attitude of one of NASA's important constituency groups, the scientific community. As far as we could tell from the numerous conversations we had with them, most scientists interested in space were opposed to the concept of the space station we were then espousing. Furthermore, it was also clear that they were more favorably disposed toward the space platform proposal advocated by the Marshall Space Flight Center.

In view of all these factors, we decided that it would be important to conduct a thorough-going study of what we should do in formulating the space station program. We asked the former NASA administrator, Dr. James C. Fletcher, to head a committee with a broad charter to develop a plan for the development of the space station that would start from first principles and would form the intellectual basis for our advocacy of the space station program. Fletcher was the right man for this assignment, not only because of his long experience in NASA, but also because of the acute technical and political judgment that he possessed. We asked him to collect a group of knowledgeable and experienced people to examine the problem in all of its aspects and then to advise us on which of the available options would be the best one to push. (A copy of my letter to Fletcher dated September 18, 1981, and a list of the people who served on Fletcher's committee are included in appendix 5.)

The Fletcher committee defined four separate options for consideration by NASA:

1. An unmanned platform, permanently maintained, for a variety of space science and applications missions.

2. A "minimum" manned platform that would allow two to three people to remain permanently in space and would be maintained and refurbished using the space shuttle.

3. An "intermediate" manned platform that would have designed into it the ability to grow in size, power, and capability.

4. A full-sized space station that could be used for servicing and supporting satellites and erecting large structures in space such as antennas and that could serve as a laboratory in space.

What the Fletcher committee did was to take the space platform favored by the Marshall Space Flight Center, call it option one; take the Space Operations Center, favored by the Johnson Space Center, call it option four; and then define two intermediate options that might be more attractive than either one of the others for various reasons. The development costs for the options defined by the committee ranged from $1.0 billion for option one to well in excess of $10.0 billion for option four. The committee carefully considered all of the options with great care, and several long meetings were held late in 1981 and early in 1982 to look at all of the facets of this complex problem. The committee firmly recommended against option one, the unmanned platform, because they felt that this did not make any technical sense as a first step toward the development of a manned system. The committee believed that only a manned space station of some type would carry with it a sufficient political imperative to interest the president and his advisors. The committee also argued that an unmanned platform would not take full advantage of the capabilities of the shuttle since the shuttle would have people on board during every flight. If the people were already there, the members felt, some kind of a habitat should be built because it would undoubtedly be desirable at some point to stay in orbit for longer periods of time than the orbital endurance of the shuttle permitted. Option two was also rejected on the ground that it would be a fundamental mistake to build something that had no potential for further growth.

This left options three and four, which thereafter became the focus of their thinking. One of the very interesting results of the Fletcher committee study was the suggestion that the large external fuel tank used by the shuttle could become one of the building blocks of the space station. This suggestion was made by one of the members of the committee (Professor James Arnold), and it is still being studied. However, the use of the external tank will probably turn out to be too expensive for the initial step. The essential outcome of the Fletcher committee's deliberations was that NASA should go ahead with the development of a manned station or, if the political and financial situations were favorable, with the full-scale space operations center.

At about the same time, the new associate deputy administrator, Philip E. Culbertson, began to collect a group of NASA people to start thinking about the space station and how the plans that were being made should be implemented. One of the people who was asked to work with Culbertson was Mr. Daniel H. Herman, who was then serving as the director of advanced planning in NASA's Planetary Program Office. Herman is a brilliant planner, and in that capacity he had been responsible for the conceptual development of many of NASA's most successful planetary programs including the Pioneer series (Jupiter and Venus), Viking (the Mars lander), and Voyager (which flew past Jupiter and Saturn and is now on the way to Uranus). Herman had (and has) close connections within the scientific community, both in this country and abroad, and he began a series of informal conversations with a number of influential scientists about the proposed space station. He reported that the scientists were concerned about the fact that there were certain scientific activities that might be harmed by the presence of people on the space station. For example, those things that required very accurate and stable pointing might be harmed by motion of the station by people moving around inside or cryogenically cooled instruments might be adversely affected by the volatile effluents from the station's life support system. It was considerations such as these that led Herman to evolve the architecture that has since become the "baseline" for the space station. His reasoning was that the station should not consist of a single unit but, rather, of several co-orbiting units that would "fly in formation," so to speak, and would each serve different functions. Thus, the space station would consist of a manned "core" that would serve as the central service station and laboratory accompanied by one or more unmanned platforms used to carry instruments that cannot be operated on the core station.

By late spring in 1982 enough work had been done on the technical concept of the space station that Beggs felt the need to create a more formal organization to carry out the work. This was a good move and an important one as well. We could not set up a formal program office because we had no approved space station program. We could, however, establish an informal task force to carry on the work and also to coordinate the work being done at the various NASA centers. This was done on May 20, 1982, and Mr. John D. Hodge was brought in from the Department of Transportation to head the task force. This turned out to be a fortunate choice. Hodge is an experienced systems engineer with a broad background in a number of different fields. He had served at the Johnson Space Center as a flight controller and was therefore thoroughly familiar both with the engineering and the operational aspects of manned space flight. The task force remained in existence until August 1984 when the formal program office for the space station program was established.

In developing our plans for the space station late in 1981, many of us felt that it would be important to use the shuttle to the utmost to demonstrate what new capabilities the presence of people in space would open up for us. The first flight of Spacelab scheduled for November 1983 would be such an opportunity and, as things turned out, it was really quite important in the final effort to push the space station program through the administration. We also felt that it would be important to demonstrate our ability to retrieve and repair satellites that had failed during their operation on orbit. As it happened, there was a good candidate for a retrieval and repair mission, the Solar Maximum Mission (SMM) satellite, or "Solar Max," as it would become known in the press two and a half years later. Solar Max was launched in 1980 and ceased to operate because of an electronics failure some months after launch. A diagnosis of the failure had revealed that the fault could probably be fixed by an operation on orbit involving extravehicular activity (EVA) by the shuttle crew. Accordingly, Milton Silveira and I proposed that the idea of a shuttle mission to rescue and fix Solar Max be implemented. This idea had been suggested earlier by a number of people. There was considerable opposition to this proposal at the time, both inside and outside of NASA. Internally, there were people at the Johnson Space Center who were opposed to performing an EVA so early in the flight program because of the inherent risks involved in these operations. There was also opposition because money would have to be taken from other programs in order to execute the mission on the required schedule. Opposition also developed in the Congress. The reprogramming of funds necessary to do the mission required approval by the House and Senate Appropriations Committees. Several members and staff people felt that the Solar Max mission was actually a demonstration of a capability more useful for the military rather than the civilian space programs, and they believed that the military should therefore pay some of the bill. Eventually, a cost-sharing arrangement was made, and approval to do the mission was secured. It was a good thing to do, as it turned out, because the successful retrieval and repair of Solar Max in April 1984 was a contributing element in persuading the Congress to approve the space station program.

At about this time one other important event occurred and this was the retirement of Dr. Christopher C. Kraft as the director of the Lyndon B. Johnson Space Center in Houston. Kraft made major contributions to the development of the important techniques of operations in space starting with the Mercury program and going through to the space shuttle. To succeed Kraft, we needed someone who had experience in space operations and who was politically astute enough to handle the difficult relationships with the Air Force and other shuttle customers as the shuttle became operational. On the advice of George Low, Mr. Gerald D. Griffin was asked to become director. Griffin was, at the time, in the private sector but had long experience in space flight operations, management, and, most important, in political relations, having served as NASA's congressional liaison director for several years. After some hesitation, he accepted the appointment and became a very effective leader of the center.

XII

Space Policy and Edwards Air Force Base – July 4, 1982

I have described in chapter 8 how the Carter administration developed its policies toward operations in space. This process would now be repeated in the new administration. The effort to secure the Reagan administration's support of the space shuttle program that led to the president's statement of November 13, 1981, has already been discussed. In addition to this, a study to formulate a space policy was now initiated. On September 8, 1981, Dr. George A. Keyworth II, the president's science advisor, distributed a memorandum that started the process. Keyworth, a brilliant, young (forty-three) physicist, who had, until recently, served as a division leader at the Los Alamos National Laboratory, was appointed as President Reagan's science advisor and as director of the Office of Science and Technology Policy (OSTP) on August 6, 1981. I had met him once or twice prior to his appointment in Washington and had developed great respect for his abilities. In his memorandum, Keyworth outlined the principal issues —just as PRM-23 had contained these in the previous administration —and provided the "terms of reference" for the study. Dr. Victor H. Reis, one of Keyworth's assistants, was appointed study director. Keyworth's memorandum was addressed to all of the federal agencies that had an interest in space operations (state, defense, OMB, the intelligence community, the Arms Control and Disarmament Agency, commerce and, of course, NASA). All of these would be represented on the "working group" that Reis was about to organize.

The interagency process to develop the Reagan administration's space policy had begun. Our first priority was clear and simple: we had to prevent the inclusion of a negative statement about new initiatives such as the one that appeared in PD-42 during the Carter administration. We did not want to be told this time that we could not start any large new programs. This was a damage-limiting objective that we felt was absolutely essential if any progress was to be made at all. Second, we wanted to have another strong statement in support of the shuttle program. This was less important once we had the president's November 13, 1981, declaration in hand. Nevertheless, we felt that it would not hurt to also have it reiterated in a policy statement. Third, if possible, we wanted an endorsement of the proposed space station program.

We were to succeed in reaching the first two objectives, but we were to fail in securing an endorsement for the space station program. There was, from the very beginning, no real controversy over the "no new programs" issue. There is no doubt that most members of the new administration were generally more in favor of the space program than the members of the previous one, and this factor helped our case. In the case of the shuttle program, the situation was more complicated. As I have already said elsewhere in these pages, there was (and still is) substantial opposition to the policy that the shuttle would be the primary U.S. launch vehicle. Both the representatives of the Defense Department and the Commerce Department (representing commercial interests) argued that it might be better to have a mixed fleet than to depend exclusively on the shuttle. This argument had (and has) considerable justification; however, we felt that it would be best to stick to NASA's traditional position on this matter, which is to defend the "shuttle only" policy. By and large, we succeeded in holding the line on this issue with only minor changes being made on the policy of the last administration.

The real fight was over the space station proposal. What we discovered was that there was strong opposition to the space station proposal from several important quarters in the administration, starting with Keyworth himself. Keyworth, in the course of this debate, took the classical position that many prominent members of the scientific community have taken on the issue of manned space flight. He said that people in space are not necessary to perform the important functions that we had in mind. On one occasion, he called the space station a "mistake" and a "step backward." Keyworth elaborated the argument by saying that we should develop the technology of unmanned spacecraft more rapidly than we have been and that we should do so in order to gain the benefits of various "spin-offs" in robotics and automation. It was in this context that he called the manned space station a "step backward" because he maintained that there would be no such "spin-offs" if we went ahead with the manned space station. When someone asked Keyworth what he thought about the Russian manned space program, he made the rather original argument that the only reason that the Russians put people in space is that their automation technology was so primitive that they had no choice but to do that. Keyworth said that we would be making a bad mistake to imitate them. Keyworth's argument was not a new one, but he made it in a more articulate and forceful way than most others who held the same view. There was also some truth in the case Keyworth was making, and we had to take this into account as well.

There was also opposition from the Department of Defense. In this case, the reason for the opposition was rather different than the one that Keyworth had used. The reasoning by the defense people was that they had no objection in principle to manned space flight, although they themselves had no requirement for putting people in space. However, they said that the amount of money that the nation was willing to spend on space operations in toto was a constant so that anything that NASA would get to spend on the space station would ultimately take away from the defense-related space programs. Finally, there was the opposition from the representatives of the OMB who made the case that a space station was too expensive and that the country could not afford to embark on such a project, given the financial situation we were facing. Both the argument made by Keyworth (manned versus automated) and the "zero-sum game" contention of the people in defense and OMB would be raised numerous times before the debate was brought to a conclusion in 1984.

There was one small concession to those who were interested in developing a space station and that was a statement in the section on research and development in which NASA was directed to "continue to explore the requirements, operational concepts and technology associated with permanent space facilities." This was at least a bow in the direction of what we thought should be done with respect to the space station, but it was clearly not enough. On balance, nevertheless, I thought that the Reagan administration's space policy was significantly better than the one that had been evolved under the Carter administration (PD-42). I was satisfied that we had done the very best we could against really formidable opposition. That we did as well as we did was largely due to the efforts of Messrs. Philip E. Culbertson and Isaac T. Gillam who were NASA's representatives on Reis's "interagency" group. (The space policy as published on July 4, 1982, is included in appendix 6.) They prevented a more restrictive statement from being written, and they therefore provided us with maneuvering room to continue the game in other arenas (the editors of *Science* magazine essentially agreed with this judgment in an article titled "Squabbling Over the Space Policy" that appeared on July 23, 1982).

On June 7, 1982, I had a long lunch meeting with Lt. Col. Gilbert D. Rye that would turn out to have important consequences in the longer term. Rye was shortly to replace Col. Mike Berta as the principal advisor for space matters on the National Security Council staff. The issue that Rye and I discussed during our long lunch meeting had to do with the final provision of the Reagan administration's space policy that dealt with the establishment of an interagency coordinating committee to oversee the implementation of the space policy. This committee, called the Senior Interagency Group (Space) or SIG (Space) had roughly the same function as the PRC (Space) had during the Carter administration, which I have already described. The question before us was to make some judgment about who should chair the SIG (Space), Keyworth or the new national security advisor, Judge William P. Clark, who had moved over from the State Department when Richard Allen was forced to resign late in 1981. Precedent was with Keyworth since his predecessor, Frank Press, had chaired the PRC (Space) in the Carter administration. I felt strongly that it would be much better to have Judge Clark (the title "Judge" went back to Clark's service as a justice on the California Supreme Court) head the SIG (Space) than Keyworth. Judge Clark is an exceedingly astute politician and was much more likely than Keyworth to see the political value that the initiation of the space station program might have for the president. I therefore urged Rye to work as hard as he could to secure the chairmanship of the SIG (Space) for Judge Clark. I also had to persuade my colleagues in NASA that this was, in fact, the best move since both Beggs and Culbertson were skeptical. I was finally able to do this and to tell Rye that it was satisfactory to NASA to have Judge Clark chair the SIG (Space).

Rye went to work and, within less than a month, he had the president's signature on the space policy document—the president signed off while on vacation at his Santa Barbara ranch for the Independence Day weekend on July 3, 1982. As it happened, our judgment that Judge Clark would be a strong supporter of the space program was critical in this case. It was Judge Clark who took the space policy paper to the president at his vacation retreat and persuaded him to sign it quickly so that he could use it for an Independence Day event that I will discuss shortly. This remarkable piece of bureaucratic legerdemain was only the first of a series that Rye would pull out of the hat in support of the space station program. I doubt very much whether the space station could ever have been pushed through the administration if it were not for the persistent and astute work of Gil Rye.

While all of this was happening, another part of the space station campaign was going on elsewhere that turned out to be much more important. I have already mentioned that the overall strategy that Beggs had conceived to push the space station through the administration was to try and interest the president directly in the space program and in the proposed space station by using the public attention that the periodic shuttle flights were attracting. A golden opportunity soon presented itself that would permit Beggs to brilliantly execute part of his strategy. As luck would have it, we discovered, early in 1982, that the landing for *Columbia*'s fourth flight (STS-4) would occur on or about July 4, 1982, at Edwards Air Force Base. Beggs reasoned that it might be possible to persuade the president to attend the landing. The president's ranch was not very far away (a short hour's helicopter flying time from Edwards), and the sight of *Columbia* coming back from earth orbit was very impressive. We also hoped that the president would take note of the half million or so people (the total people count after the fact turned out to be about 525,000) that would be there in the desert to watch *Columbia* come in along with the president. Beggs reasoned that, as a master politician, the president would draw the right conclusion from what he would see and get a better feeling for the political value of the space program.

In the early months of 1982, Jim Beggs worked hard to establish strong contacts with various members of the president's staff in the White House. As a member of the SIG (Space), he would lead the interagency arguments, but he recognized that this was not nearly enough and that routes of access to the president other than through the science advisor and the national security advisor would have to be found. The two most important people that Beggs had contacted were Mr. Michael K. Deaver and Mr. Craig Fuller. Deaver is a longtime friend of President Reagan's and is the deputy chief of staff at the White House, among whose responsibilities it was to see to it that the president's schedule was properly organized. When the proposal that the president attend the *Columbia*'s landing on July 4, 1982, was made, Deaver liked the idea. This was an important factor in organizing the event. Fuller was (and is) the youthful secretary of the cabinet, and he also became very intrigued with the possibilities inherent in the space program. Fuller's primary interest is in the possible commercial applications of space operations, and he would, in the coming months, play a very important role in persuading the president to adopt the space station program. Fuller is very bright and articulate and, as the secretary of the cabinet, he has strong influence over the agenda and the scheduling of cabinet meetings, which would be very valuable in developing the timing and strategy for the various steps that were necessary to push the space station program through the administration. Beggs had done well to develop these contacts, and they were now to pay off in a handsome way. (As in the last administration, I was not directly involved in the policy-making process since I was not a member of the SIG (Space), nor was I directly involved in the development of other relationships with the White House. I was close enough, however, to see what was happening.)

In developing the plans for the president's visit to Edwards Air Force Base on July 4, 1982, Jim Fanseen would again play a critical role in arranging the event. This was an extremely difficult and complex task, and the fact that things went as well as they did is very much due to his excellent work.

On June 9, 1982, Jim Beggs left for a two-week vacation in Europe, and he asked me to continue to ride herd on the planning process for the president's July 4 visit to Edwards Air Force Base that he had initiated. The most important matter before us at the time was to prepare the speech that the president would make after *Columbia*'s successful touchdown. On June 17, 1982, there was a meeting at the White House (in Craig Fuller's office) to discuss the information briefings we would provide for the president before the July 4 event and to discuss the speech that he would be making. Craig Fuller, Richard

Darman, a senior policy advisor to the president, Gil Rye, and I were at the meeting. President Reagan uses videotapes very extensively to gather the information he needs to perform his job. Accordingly, we had prepared a short videotape describing NASA's activities so that the president would have the necessary background information. We also discussed the president's speech for the occasion. I had prepared a draft of the speech, which I brought along to the meeting for discussion. (The draft speech is included in appendix 7.)

The major issue we discussed had to do with whether we should include in the speech an announcement that the president would make committing the nation to the construction of a space station. I had put such an announcement in the draft of the speech that I had prepared, but I put the statement in brackets knowing that it would be controversial. All of the people at the meeting were strong supporters of the space program and were oriented toward the eventual initiation of the space station program. I argued that it might be a good idea to include it now. There was much talk in the press about getting a space station in the NASA program at this time (after all, Beggs and I had been making the suggestion for over a year by now), and I argued that people would be disappointed if the president did not say something about this matter. In spite of my arguments, the consensus of opinion at the meeting was that we should not include the announcement of a space station in this speech. The argument was made that the people in OMB and OSTP would both strongly oppose such an announcement and would undoubtedly win the debate to strike the paragraph from the speech. I had to agree with this assessment and reluctantly deleted the statement. There was, however, one other point on which I did insist. I felt that the sentence "The Space Shuttle gives us the means for establishing the permanent presence of mankind in space" or something like it absolutely had to be included in the president's speech. There was general agreement on this point, and all present agreed that each would help to retain the sentence. My belief was that if we succeeded in doing this, we could at least claim that we had achieved a step in the right direction since "the permanent presence of mankind in space" could clearly be interpreted as working toward a space station.

This phrase soon became controversial. On June 30, 1982, I had a call from Gil Rye informing me that both the people in OMB and in Keyworth's office were anxious to delete the words "permanent presence in space" from the speech since they realized what we would do once it was on the record. I asked Rye to mobilize our friends on the White House staff to see what they could do and also talked to Beggs about the problem. The current plan was that he would go to Edwards to be with the president for *Columbia*'s landing and that he would make certain to get a look at the speech before the president delivered it. If the "permanent presence" phrase was not in the speech, he would do whatever he could to get it restored.

As was my custom, I went to Houston for the landing. This time my wife went along for the holiday weekend to participate in the festivities that would be part of the landing. We could watch things from the Mission Control Center—I was at my usual console and she was in the viewing room. The television monitors showed a beautiful day out in California—a perfect July 4. *Columbia* was scheduled to touch down a few minutes after 9:00 A.M., California time. I followed the landing sequence, and as scheduled the entry sequence was properly executed. *Columbia* became visible on the screen, and in a few minutes the great bird touched down on runway twenty-two for a perfect landing. The President and Mrs. Reagan drove out to welcome the crew (mission commander T. K. Mattingly and pilot Henry Hartsfield) as they came off the airplane. They walked around *Columbia* on a short inspection trip, and then they departed for the speech-making ceremonies. There was the usual delay with the crew receiving a quick medical examination. I also learned later that there had been a short ceremony during which the president gave Mattingly and Hartsfield their Distinguished Service Medals. He also pinned the third star on Jim Abrahamson's shoulder since Abrahamson had recently been nominated for promotion to the rank of lieutenant general. (The last part of the ceremony was a complete surprise to Abrahamson and had been arranged by Jim Fanseen.)

Then the president made the speech. Jim Beggs introduced the president and used the line from the president's inaugural speech about "great nations dreaming great dreams." It was well done. The president started to make his speech, and after the first few words it was clear that the speech was rather different than the draft I had prepared for Craig Fuller (see appendix 7). However, about halfway through the speech, the president clearly and firmly said that it was now our intention to "look aggressively to the future by demonstrating the potential of the shuttle and establishing a more permanent presence in space." So, the important statement was there, and we had succeeded in our effort to get it on the record. The rest of the president's speech was very good. It was somewhat longer than the NASA draft, and it was more detailed in delineating the provisions of the space policy statement that the president had signed on the previous day. (The text of the speech is included in appendix 8.)

The ceremony at Edwards ended in a really impressive way. The second Orbiter, *Challenger*, had recently been rolled out and was now ready to be moved to the Kennedy Space Center. *Challenger* had been put on top of the Boeing 747 shuttle carrier aircraft and just after the president finished his speech, the Boeing with *Challenger* on its back flew over the reviewing stand on the first leg of its trip to Florida. It was a grand way to end the ceremony.

For us in Houston, the ceremonies were not yet complete. *Columbia's* crew would be returning later on in the day and would be landing at Ellington Air Force Base a few miles away from the Johnson Space Center. We also had planned to have *Challenger* and her Boeing 747 carrier aircraft land at Ellington for a refueling stop. When it became known that *Challenger* would be at Ellington, a crowd of some tens of thousands of people quickly collected, and we had quite an impromptu celebration. Both Ken Mattingly and Henry Hartsfield made short and eloquent speeches and received a hero's welcome from the crowd. I also made a short speech in which I mentioned the president's "permanent presence" statement and also talked about the space station. If the project were approved, I said that the job would be done in Houston. Naturally, I was roundly cheered when I said all of this because I was preaching to the choir. It was a really great event for all concerned, and I felt honored and pleased to participate.

The events of July 4, 1982, confirmed what we learned during the president's visit to the Mission Control Center at the Johnson Space Center the previous November. The president had a very definite interest in NASA and the space program and had a deep and knowledgeable appreciation for the political value of our space operations. This was a really important conclusion, and it shaped our actions in the coming months. I know that what I have written in this chapter could be interpreted by some as simple manipulation of the president and his staff. As a participant in the events I have recounted, I can attest to the fact with absolute certainty that neither the president nor his principal assistants were the kind of people who could be manipulated by anyone. The truth is rather different. What we did, and this was the central feature of Jim Beggs's strategy, was to expose the president and his aides to the space program, and they then drew their own conclusions.

At about the same time that the president made his visit to Edwards Air Force Base to watch the touchdown of *Columbia* from her fourth flight, we were putting the Fiscal Year 1984 budget together. The principal issue was whether we should ask for a large appropriation for the space station or whether we should concentrate on adding a fifth Orbiter to the shuttle fleet. (As I have already said, see chapter 6, the original plan for the shuttle fleet called for five Orbiters.) After some thorough discussions, we elected to concentrate on the fifth Orbiter and to hold back on the space station for the time being. This proved to be an excellent decision on Jim Beggs's part, although I have to confess that I did not think so at the time. What we finally did was to put in a modest amount of money (about $30 million) for various space station studies and technology development efforts and to ask for the fifth Orbiter. We succeeded in securing some support for the space station, and while we failed to gain approval to build a fifth Orbiter, we did succeed in arranging a compromise that permitted us to keep the shuttle Orbiter production line going to produce "structural spares" that could be used to assemble a fifth Orbiter if that became necessary. The outcome of the Fiscal Year 1984 budget negotiations were generally favorable to NASA, and we prepared to push for the space station next year.

Having made the decision to not push for the space station in the Fiscal Year 1984 budget process, we had to start making the plans for what to do in Fiscal Year 1985. Because of the way the timing of the budget process works, the Fiscal Year 1985 budget would be put together in the spring of 1983 and would be submitted by NASA to the OMB in September 1983. Since 1984 was an election year with all of the attendant uncertainties, we reasoned that the Fiscal Year 1985 budget process would be the last appropriate time for several years to push the space station project through. We, therefore, mobilized ourselves for a major effort that consisted essentially of two activities. First, we continued to work with the senior members of the president's staff to maintain and enhance the president's interest in the space program. What we

were looking for were opportunities for the president to make a commitment to the space station in an appropriate major speech. Second, we
had to start the interagency review process that would be required to get
a major initiative like the space station. We did not believe that the
outcome of the interagency process would be very positive so that a
presidential commitment would be most important in short-circuiting
the arguments. The interagency process would be carried out through
the SIG (Space), which had all of the interested parties as members. On
October 1, 1982, there was a routine meeting of the SIG (Space) with
Mr. Robert (Bud) McFarlane in the chair. McFarlane was Judge Clark's
deputy and ran the day-to-day operations of the National Security
Council. I sat in for Jim Beggs at the meeting since it was only an
information rather than a decision meeting. I gave a short report on the
shuttle program and what we had learned in the first four flights. I also
made a short statement about the space station, which was, I believe,
the first time this subject—which would soon become very controversial
—was mentioned at a SIG (Space) meeting.

The senior members of the president's staff, Mr. Edwin Meese III,
Mr. Craig Fuller, the director of OMB, and several others had adopted
the idea of making personal visits to the various large federal agencies
in order to learn firsthand about some of the things that were going on.
A visit to NASA was scheduled for December 2, 1982, which was attended
by Meese, Fuller, and Mr. Joseph Wright, the new deputy director of
OMB. We had another opportunity at this meeting to make our case for
the space station to this group, and we took it. I gave a short talk about
the history of NASA and where we had been, and Jim Beggs gave a really
brilliant talk about what we had in mind for the future. He mixed the
technical, political, and commercial aspects of our space activities in
exactly the right proportions for this group and made an excellent
impression. On leaving this meeting, I felt that we had some strong
friends among the president's senior staff people. Craig Fuller was
especially positive, and he later on was to provide important support at
critical moments.

In August 1982 there was an international Congress on Space sponsored by the United Nations called UNISPACE 82. The Congress was
held in Vienna, Austria, and along with the speeches, there would also
be an exhibition. Jim Beggs headed the U.S. delegation, and he was
assisted by Ambassador Gerald Helman of the State Department. Mrs.

Joan Clark, wife of the president's national security advisor, was also a member of the U.S. delegation. I went along as a special advisor to the delegation, but the real purpose for my going had to do with a series of television and radio interviews and debates that I had been invited to participate in by the Austrian television network. Although I was not born in Vienna, I lived there as a young boy, and I still speak German well enough to pass muster on German-speaking television programs. I discovered that Mrs. Clark had been born in the German-speaking region of Czechoslovakia and that she also spoke the language well enough to appear on television interviews. After a few television appearances, we began to call ourselves the "Clark-Mark Show." What was important about all of this is that several of these interviews were staged at the U.S. exhibit at which there were a number of nice models of space stations. We thus had a good opportunity to acquaint the visitors to the exhibit and the European television audience with what we had in mind.

During the UNISPACE 82 meeting, I also had the opportunity to speak informally with some of my German friends including, among others, Hans Hoffmann of the MBB-ERNO organization in Bremen. The topic of our discussion concerned the space station and how Germany and other European countries could participate in a space station program. By this time, I felt that the chances of actually initiating a space station program were good enough that such conversations would be useful. Hoffmann's organization had a very special interest in the space station since they were the primary contractors for the Spacelab program of the European Space Agency (ESA). (The Spacelab is a manned module that fits in the payload bay of the shuttle and is used for the conduct of onboard experiments. The Spacelab flew for the first time in November 1983 and on the STS-9 mission.) Hoffmann was extremely interested in the proposition of some collaboration on the space station program if there was to be one and promised to talk about it to others in Europe who might be interested in participating. At this time, I did not have a strong opinion about international collaboration on the space station program because I felt that this was a policy matter somewhat out of my province. However, I felt that it might be useful to make some initial contacts, and these turned out to be most important later on.

XIII

The *Enterprise* in Europe— May–June 1983

Every second year (the odd-numbered years), there is a large international air show in Paris. Many nations bring their newest aircraft and other exhibits to Paris for this event, and the larger countries each have fairly elaborate "pavilions" of their own in which pictures, movies, and exhibits are shown. Early in 1983 the question of what the United States should do at the air show was discussed and Mr. Robert F. Allnutt, a senior advisor to the NASA administrator, suggested that it might be a good idea to make the space shuttle the centerpiece of the U.S. exhibit. We had flown the shuttle successfully a number of times, and the flights had attracted considerable attention. The reasoning was that we could attract favorable international public attention to the space program by taking one of the shuttle aircraft over to Paris and putting it on exhibition at the air show scheduled to start in May 1983. Hopefully, this public attention would be duly noted by senior members of the administration. To implement this suggestion, the proposal was made to use the *Enterprise* for this purpose. (I have already mentioned *Enterprise* in connection with the Approach and Landing Tests conducted in 1977. See chapter 7.) *Enterprise* would be flown across the Atlantic on the back of her Boeing 747 carrier aircraft and would be put on exhibition at Le Bourget, the airfield just outside Paris, where the Paris Air Show is held.

We all thought that this idea was an excellent one and that we should do whatever we could to make it happen. However, there were some very real problems. There was, and, unfortunately, still is today, only one Boeing 747 that is suitably modified to be able to carry the shuttle Orbiter. Taking the airplane across to Europe would entail some risk of loss of the aircraft, since we depended on the shuttle carrier aircraft to ferry the shuttle from the landing site (wherever it happened to be—we had landed both at Edwards Air Force Base in California and at the Northrop strip in New Mexico) back to the launch pad at the Kennedy Space Center in Florida. If we lost the Boeing 747 on the trip to Europe, this could easily cause very serious flight schedule problems for the shuttle program because we could then only land at the Kennedy Space Center until a new carrier aircraft was built—a process that might take up to two years. The loss of the *Enterprise* could also be serious. Although *Enterprise* was not designed to ever fly in space, she is used as a full-scale mockup vehicle for many purposes. The most important of these uses will be in the activation of the west coast launch site in 1985 during which *Enterprise* will be used for the "form-fit-and-function" checkout of the launch pad. Before making the decision to send *Enterprise* to Europe, we had to assess these risks.

In addition to the technical risks that we would have to accept, there was also the problem of terrorists. The *Enterprise* and her carrier aircraft would make a tempting target for any group that was interested in attracting international attention, and the possibility of a terrorist attack had to be factored into our thinking. We asked the State Department and various security agencies to make a risk assessment for us on the terrorist question. We combined these estimates with the technical risks we knew were there and, after a thorough review, decided that the risk of sending *Enterprise* to Europe was acceptable, and so a decision was made to go ahead with the plan. As was to be expected, the word that we were planning to send *Enterprise* to the Paris Air Show soon leaked out, and we had diplomatic representations from other European nations to have *Enterprise* make visits to them as well. Accordingly, a more elaborate trip was arranged during which *Enterprise* would be in Paris for the most important portion of the 1983 air show, but would also visit London, Bonn, and Rome, the capitals of our three major European allies.

It was determined that both Jim Beggs and I would go to Europe during the time that *Enterprise* would be there. In addition to talking about the space shuttle, we would also take the opportunity to continue the process that we had started the year before at UNISPACE 82 to sound out the Europeans about their interest in a space station program. I decided that I would stay, as much as possible, with the *Enterprise* and travel with the crew on the shuttle carrier aircraft. My wife, Dr. Marion T. Mark, would accompany me on the trip. The Beggses would spend more time in Paris so that Jim could talk with the people at the European Space Agency (ESA) headquarters, and I would spend most of my time in Germany making contact with the political and industrial people there.

On May 19, 1983, my wife and I took the night commercial flight to London. We arrived early in the morning and drove out to RAF Fairford, which is an RAF base where we were to meet the *Enterprise*. The events at Fairford were a foretaste of what was to come. The stop at Fairford was intended solely as a refueling stop and no prior announcement of the arrival of *Enterprise* and her carrier aircraft had been made. This was done deliberately since the first official landing in Europe was scheduled to take place in Germany as part of the celebration to mark the 300th anniversary of the first German settlement in North America. In spite of the fact that there was no announcement, there were over 30,000 people at Fairford to watch *Enterprise* land. In addition, a number of dignitaries, including the lord lieutenant of the county (Gloucestershire), had also assembled so that we had some unexpected representational duties to perform. It was an impressive sight to see the great aircraft with the gleaming white *Enterprise* sitting on the back flying over the airfield. The airplane made one low pass over the runway and a great cheer went up from the crowd. Somehow, the sight stirred some very deep emotions in people who saw it, and we were to see a repetition of this scene wherever we went in Europe.

We boarded the Boeing 747 shuttle carrier aircraft after a short wait for the refueling and set our course for Germany. The mission was headed by Col. Richard L. (Larry) Griffin who was at the time serving on a detail from the Air Force to the Johnson Space Center. The three pilots were Mr. Joe Algranti (the chief pilot at the Johnson Space Center), Mr. Dick Scobee (an astronaut), and Mr. Tom McMurtry. In addition, there were several other members of the crew, including Ms. Judy Elam of the Johnson Space Center who handled the security arrangements. We arrived over Koblenz, Germany, on the Rhine at about 5:00 P.M. and flew down the Rhine to land at the Bonn-Cologne airport shortly after 5:00 P.M., on May 20, 1983. Even though the weather was bad, there were large crowds on both sides of the river, and about 20,000 people were on hand to greet us. The German minister of science and technology, Dr. Heinz Riesenhuber, who would later become one of the strongest proponents of European participation in the space station program, was the senior German official there to greet us. He was accompanied by Drs. Wolfgang Finke and Herman Strub, two senior officials of the ministry.

The *Enterprise* attracted much attention in Germany. During the two days we remained at Bonn-Cologne (May 21 and 22), about a quarter of a million people came to see the *Enterprise* perched on top of the shuttle carrier aircraft parked on one of the airport ramps. We were overwhelmed by the numbers since we did not have nearly enough people to provide the tours and the lectures we had in mind. This was true even though Col. and Mrs. Karol Bobko joined us in Bonn. (Bobko was the pilot on the STS-6 mission.) There was no question from what we saw at Bonn that our mission to Europe would be a great success. The shuttle had caught people's imagination, and great numbers of people took the opportunity to go and take a look for themselves. We hoped that this public interest could be translated into the appropriate political support when we needed to have it.

After two days in Bonn, we flew *Enterprise* to Paris to participate in the opening ceremonies of the air show. Once again, what we had seen in Germany would be repeated. The presence of the *Enterprise* drew crowds and caused much comment—almost all of it favorable—in the press and on television. There was no doubt at all that *Enterprise* was the main attraction at the air show. I met Jim Beggs in Paris, and we then began the talks that we had scheduled with the leaders of the European space program. We (my wife and I) returned to Germany on May 29, 1983, driving through northern France and Belgium for a short vacation from the chore of exhibiting *Enterprise*. During the next few days, I had several meetings with senior German officials of the Ministry of Science and Technology and also the DFVLR (Deutscher Forschungs und Versuchverein für Luft und Raumfahrt), which is roughly the equivalent of NASA in Germany since it operates the technical establishments.

Several meetings with Drs. Finke and Strub and the members of their staffs at the Ministry of Science and Technology were arranged. I described to them in detail our thinking about the space station and what I thought could and should be done. In order to make a credible case that the United States might actually make the commitment to build a space station in the coming year, I used the argument that the politics of the presidential campaign would help move the administration to adopt a space station program. At the time, Senator John Glenn (D., Ohio) had just announced his candidacy for the Democratic presidential nomination, and he was making a strong showing in the public opinion polls. Glenn had been arguing for a strong space program for a long time. As a former astronaut and a distinguished political leader, he understood the technical importance and the political value of such a program. He had also endorsed the construction of a space station as the next step after the shuttle program was completed. I told the Germans that I thought a strong showing by Senator Glenn would perhaps be the trigger that would start the space station program. I was wrong in this judgment as things turned out, and I don't know if the Germans believed me, but it is probably not important. Very often, the right things are done for the wrong reasons!

What is probably more important, I also spent some time listening to the Germans rather than talking. On a visit to the DFVLR laboratory as a guest of Professor Jordan, the director, I learned something about what the Germans were doing in the field of materials processing in space. This is very definitely one of the really important areas for further work on Spacelab and, ultimately, on the space station. It was impressively clear that the work in Germany was pointed in the direction that would be very important if a space station program were initiated. I also visited Hans Hoffmann and his collaborators at the ERNO facility in Bremen. They had built the Spacelab modules and were, therefore, most interested in applying what they had learned about spacecraft construction and life support systems during the Spacelab program to the space station. They clearly also had an extremely competent group of people who could make most important contributions. During the next few days, I also visited the Messerschmitt-Boelkov-Blohm (MBB) facility near Munich and the Dornier factory in Friedrichshafen. In all of these places, I again saw good facilities, excellent people, and a clear and strong commitment to the space program. The sum of my discussions with the Germans was that they were most interested in participating with us in a space station program and that they would want to organize this support somehow through the European Space Agency (ESA).

We returned to Paris in time to fly to England with the *Enterprise*.

June 5, 1983, was the last day of the Paris Air Show and after the closing ceremony, we took off from Le Bourget shortly after noon. We had arranged to fly over the Low Countries before crossing over to England. It was a memorable flight. We started with Charleroi, then Brussels and Antwerp, with Rotterdam and Amsterdam toward the end of the circuit. Everywhere, there were thousands of people on the ground looking at us as we flew by at 3,000 feet. I had never seen anything like it—we must have made an awesome appearance. We crossed over to England at around 3:00 P.M. Our flight plan called for us to fly up the Thames River to Windsor Castle at about 3,000 feet, make a circle around Windsor Castle and then fly up to Stansted Airfield (about forty-five miles north of London) for our landing in England. Unfortunately, the weather was overcast so we could not execute the original plan. We did fly over to Windsor and dropped down to 3,000 feet to circle the castle. Since it looked like the weather was lifting, Joe Algranti called the Heathrow Tower and asked to fly down the Thames at 3,000 feet. Permission was granted and the next twenty minutes were absolutely fascinating. It was Sunday afternoon and many thousands of people were lining the river watching for the *Enterprise* to pass overhead. The crowds were enthusiastic—even at that altitude we could see them cheering and waving. Then, in the course of three minutes, we flew over the Parliament at Westminster, the Tower, and the famous observatory at Greenwich. There, in a few minutes, we saw the important places where our political and technical and scientific heritage arose. It was an emotional experience.

If overflying London was an experience, then our landing at Stansted was completely overwhelming. Nowhere else in Europe did we get the kind of reception we experienced. There were crowds elsewhere, but in this place the enthusiasm was hard to believe. The people cheered and held their hands up making the "V" for victory sign and many had tears streaming down their faces. I cannot explain why this happened, I can only record it. We were welcomed at Stansted by U.S. Ambassador and Mrs. John Louis and Mr. and Mrs. Patrick Jenkin (Mr. Jenkin was the minister of industry). I made a short speech in which I talked about the close relationship between the British and the American scientific heritage. I pointed out that two of the four sister ships of *Enterprise, Discovery* and *Challenger* were named after British research vessels, illustrating the close relationship between British and American traditions in exploration. This remark was much appreciated, and I received a good round of applause and much cheering. It was a memorable occasion.

We spent a few days in Britain, and I had some conversations with people at the Royal Aircraft Establishment (Farnborough) and also at the Defense Ministry in London about British participation in a possible space station program. There was much interest, but since Britain has a substantially smaller space program than Germany, these conversations were much less specific.

Upon our return home, we reviewed the results of the visit of *Enterprise* to Europe. There is no doubt that we attracted much public attention. We estimated that well over two million people saw the *Enterprise* either on the ground or during her overflights of various parts of Europe. (*Enterprise* also visited Rome in addition to Paris, Bonn-Cologne, and London.) We also sensed that we had made an impression on a number of political leaders and that this eventually could be useful if the administration chose to make the space station program an international one. We concluded that the most important effect of the visit of *Enterprise* to Europe was to keep the international option for the space station open.

XIV The Final Push

Having made the decision to include the space station in the Fiscal Year 1985 budget, we now had to go through the process of developing an interagency position on the space station proposal. I had some hopes that we would be able to gain support for the space station in this process from other departments in the federal government. Jim Beggs was less optimistic, but he had a better view than I of the situation in the White House. He believed from the beginning that the essential matter was to have the president solidly on our side and that if we could hold his interest, then, eventually, he would make the decision to go ahead. This did not mean that Beggs did not also have moments of doubt—we all did at various times along the way—and I will mention some of these in due course.

On February 9, 1983, I presented a briefing to the Defense Science Board (chaired by Mr. Norman Augustine of the Martin Marietta Corporation) on the shuttle operations that we had carried out so far and also on our plans for the space station. The reception I received was very critical, with few board members saying anything favorable about our plans. They were impressed by the technical success that we had with the shuttle, but they were skeptical of our ability to turn that into an operational or an economic success. There was (and still is) some truth to the argument that NASA must still demonstrate that the shuttle will be operationally as useful as we have advertised. The attitude toward the space station was that this was undoubtedly something that should eventually be done but that it was premature and not well justified. I was depressed by what I heard but also knew that the views of this very influential committee could not be ignored.

A month later (March 12, 1983) Gil Rye came over for lunch to plan the procedures for starting the "interagency process" on the space station. There were two parts of his plan: One was to get a meeting with the president for Jim Beggs so that he could describe our plans for the space station to the president in person. The other was to develop a paper stating the "terms of reference" that would be used to guide the space station study that would be carried out before the president made his final decision in the fall. Based on what I had heard so far, I was not as optimistic about this process as I had been, but I did feel that we would have to do a very thorough job.

On March 28, 1983, there was a meeting of the SIG (Space) in the White House Situation Room at which the terms of reference for the study on the space station, as well as some other studies, were discussed. It was an interesting meeting, and I was pleased to be invited to attend even though I was not a member of the SIG (Space). Judge Clark was in the chair; Mr. Paul Thayer, the deputy secretary of defense, spoke for the Defense Department; and Mr. James Malone, assistant secretary of state for oceans and international environmental and scientific affairs spoke for the State Department; and the other interested departments were also represented by high level officials. Although this was not a decision meeting, it was clear that we would have problems. Several of the people around the table made negative comments about NASA's plans. Jim Beggs and I were particularly concerned about trying to influence Paul Thayer since he had been the chief executive officer of an aerospace corporation (LTV) and could thus be expected to be more sympathetic to what we were trying to do. Beggs did not make an issue of it at this meeting, but we realized now more than before that we would have problems. A few days after this meeting, the terms of reference were issued and a working group was established to produce an options paper from which the president would eventually get the information to make a decision on the space station (see appendix 9). It is significant that the president himself signed the terms of reference. He did so at the urging of Judge Clark and Gil Rye. His signature on the document made the subsequent debate easier for those of us who were strong proponents of the space station.

We now had to put together the strongest team we could muster to represent NASA on the working group and to conduct the day-to-day tactical moves that would be required in dealing with the interagency process. Fortunately, we had some very experienced people available to organize this effort and then to participate as the major actors. Phil Culbertson had spent almost twenty years as a senior official at NASA headquarters and had worked on interagency matters before. During the Carter administration, for example, Culbertson was NASA's representative to the antisatellite weapon limitation talks so that he had considerable experience in dealing with the bureaucracy in the State and Defense departments. Another very important person was Mr. Norman Terrell, who had been appointed as NASA's associate administrator for policy the previous December. Terrell started his career as a foreign service officer and had served as NASA's director of international affairs during the Carter administration. At the beginning of the Reagan administration, Terrell was nominated to become an assistant director of the U.S. Arms Control and Disarmament Agency (ACDA) since he had wide experience in arms control matters as well. He was strongly supported for this post by Professor Eugene Rostow, the director of ACDA. The nomination ran into political difficulties in the U.S. Senate because of a dispute between Rostow and Senator Jesse Helms, and it was eventually withdrawn. Because of Terrell's important service at ACDA during his interim appointment, NASA was asked to find a post for him. I initially opposed Terrell's appointment because I felt that NASA should not be used for the purpose of finding positions for people with political problems. As it happened, I was wrong. Terrell turned out to be one of the principal strategists in pushing the space station through the bureaucracy. He was extremely effective in doing his job.

In addition, there were a number of other people who made major contributions. There was John Hodge, the chairman of the Space Station Task Force, who led the early definition efforts that were so important in persuading people to accept the idea of a space station, and several others, including Luther Powell, Dan Herman, and Terry Finn, were significant contributors as well. Perhaps, however, the most important member of the space station strategy group was Mrs. Margaret Finarelli, who would become NASA's representative on the working group that the SIG (Space) had established. This was perhaps the most difficult and contentious of all the jobs that had to be done. Finarelli was (and is) a staff member of NASA's International Affairs Office, and she had wide experience in Washington to bring to her job. She had served for a time in the intelligence community and was, therefore, familiar with some of the arguments that she would hear from the other members of the working group. Finarelli became our frontline soldier in the whole effort, and she did her job with grace, intelligence, and good humor.

We had a number of meetings to look at our overall strategy for the interagency process at this time. Our objective would be, of course, to try and get the agencies involved to support the space station proposal so that we could present a position paper to the president that would contain a consensus position in favor of going ahead with the program we had proposed. Failing that, we would conduct a damage-limiting operation in the working group and would rely on Jim Beggs to continue his efforts to persuade senior members of the White House staff and the president himself that going ahead with the space station was in the national interest. An opportunity for Beggs to do this would come soon because the meeting with the president that Gil Rye had talked of was now scheduled for April 7, 1983. I did not go to this meeting since it was meant to be as small as possible, but Beggs was quite optimistic about the future when he returned to tell us what had happened. He told us that the president was very interested in the general direction of the space program and in the space station in particular. Beggs also told us that we had some very strong support among members of the White House staff. So far, so good.

At about this time, I started to hold a number of talks with people who were likely to be important in developing a consensus of opinion on the space station. One of these was my old friend, Dr. Eugene G. Fubini, who was then serving as the vice-chairman of the Defense Science Board. I went to visit Fubini on April 16, 1983, and had a wide-ranging discussion with him on the subject of the space station. Fubini was opposed to our space station proposal because he felt that people in space were a liability rather than an asset. His primary case was that whatever should be done in space can be done with automated machines more effectively and less expensively than by using people. He also said that from a military viewpoint, the presence of people in space created a vulnerability that we did not have to create. These were powerful arguments, and we would hear them again from various sources in the coming months. The nub of the question here is, of course, whether machines really can "do everything that has to be done." If this case is pushed far enough, then the machines that have to be built to reproduce what the human mind and the human hand can do become either prohibitively expensive or impossible to build altogether. Unfortunately, it is not really possible to make a reliable, quantitative analysis of this question because it is always subject to a judgment of "what has to be done." Fubini and I agreed on this point, but we could not really agree on anything else. Fubini's argument that our space program should be driven by quantitative requirements is one that would permit such a quantitative analysis on a case-by-case basis, and he maintained that this is how our space program should be organized. The only counter to this position is that people will go into space for nonquantifiable reasons and that Fubini's argument was therefore somewhat beside the point. At the end of our conversation, Fubini told me that he had opposed the space shuttle program in 1972 for the same reason when he was a member of President Nixon's Science Advisory Committee. He said that I should not worry—he had lost the argument in 1972 and was likely to lose it again this time. Somehow, I thought this was small comfort indeed.

A few months later, I would hear the same argument from Dr. Wallace G. Berger, who was the senior staff member of the Senate appropriations subcommittee that handles the NASA budget. Berger was at the time a longtime and highly respected member of the congressional staff. He is a brilliant person of great intellectual capacity, and he wielded very considerable influence. I had a long lunch with Berger on July 11, 1983, during which he put me on notice that he would urge the Senate committee to oppose our proposal for a permanently manned space station. He told me that he would recommend a statement in NASA's Fiscal Year 1984 Appropriations Bill that would ask NASA to look carefully at building an automated space platform before we proposed a permanently manned space station. Berger asked for the same kind of quantitative trade-off study that Fubini had mentioned, and we had almost the same discussion that I had conducted earlier with Fubini. Berger added another argument that would also become important in the coming months. He pointed out that if we went ahead with a program to develop an automated space station instead of a manned one, there would be more spin-offs in the area of robotics that would ultimately have important consequences for American industry. Berger made a vigorous argument to the effect that this would be a more productive way to spend tax dollars than to go with the NASA proposal as he understood it. I countered by saying that we had put a great many automated systems on all of our manned spacecraft over the years and that I felt it would be very difficult to measure the "spin-offs" from an unmanned program as opposed to a manned one. We did not agree on this point, and once again I was left with a feeling that we would be facing very strong opposition to our space station proposal in the Congress even if we succeeded in persuading the administration to adopt it.

A few days later (July 17, 1983), I had lunch at the Cosmos Club with my old mentor, Professor Edward Teller, and I heard another view that would eventually be used by the opponents of the space station proposal. Teller had been named to the White House Science Council that the president's science advisor, Dr. George A. Keyworth II, had established, and so his views on what NASA should be doing could be important. Teller started our conversation by saying that his major objective was to go back to the moon and to construct a permanently manned lunar base some time before the year 2000. He was interested in this for intrinsic scientific reasons, but he was also afraid that if the Russians established such a base before we did, we would suffer—at the very least—a propaganda defeat. At worst, he said, a Russian lunar base would give them capabilities that might have some serious national security implications as well. He was not specific about what problems such a Russian lunar base would cause for us, but he said that such things always seem to have a way of impinging on national security, and he felt that this one would be no different.

I had to take what Teller said seriously because he has a long and distinguished track record about judging things of this kind. I replied by pointing out to him that the space station was a necessary and essential step toward the development of a lunar base. We would not be able to put a base on the moon without having an operations base in earth orbit first. Teller, who knew the numbers as well as I did, responded by conceding that such a station would be necessary, but that it would not have to be permanently manned. Furthermore, he said that he was not at all sure that the space station we had in mind could be easily adapted to serve as a staging base to go back to the moon. He wanted to know whether we had factored the eventual establishment of a lunar base into our space station plans and whether this led to a permanently manned space station. We had, of course, put the staging base consider-

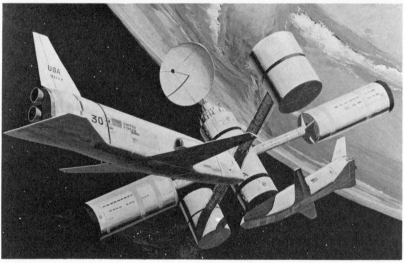

Two earlier concepts (late 1960s) of the space shuttle and space station.

A design configuration for the space station done in 1982.

ation in our plans, but we did not consider a temporarily manned space station as Teller seemed to be proposing. (See the recommendations of the Fletcher committee mentioned in chapter 9.)

Here, then, was another argument that we would be facing in the coming months, but it would come from a different direction than the ones made by Fubini and Berger. There would be people who would say that the space station is not enough and that we should develop a much more ambitious plan. We should set a long-term goal—such as establishing a permanent base on the moon or perhaps mounting a manned expedition to Mars—and that the space station should be designed simply as a step toward the more ambitious goal. This line of reasoning had the potential of stopping NASA's plans just as effectively as the negative argument because the establishment of a more ambitious goal of this kind would take several years of study before it could be approved by the administration. I made this point in the conversation with Teller, but I cannot say that I persuaded him. (NASA did, of course, have study efforts going to define what could be done to establish a lunar base. These were being carried out at the Johnson Space Center under the direction of Drs. Michael Duke and Wendell Mendell. While these studies were extremely valuable, they were very small, certainly not on the scale of what Teller had in mind.)

I had one other meeting during this period at which I learned more about the opposition that we would be facing in the fall. This was a long lunch with Mr. Frederick N. Khedouri, the associate director of OMB for natural resources energy and science. Khedouri is an astute young lawyer who served on the staff of the OMB director, David A. Stockman, when the latter was serving in the House of Representatives. He joined the OMB when Stockman became the director early in 1981. Our lunch meeting ranged over a number of topics including some national security related problems that were current at the time. Toward the end of our lunch, Khedouri asked me about the space station. Although our Fiscal Year 1985 budget would not be submitted for another month, I saw no harm in telling Khedouri what it would contain. (By this time, it was general knowledge that NASA would try to persuade the administration to approve a space station program for Fiscal Year 1985.) Khedouri was concerned about the analyses we would present to justify the space station. As far as he could tell, he told me, there was essentially no analytical justification based on the requirements of our space program for the space station program we were proposing. He said that the administration was facing very serious financial difficulties, and he said only those programs that could withstand the most rigorous analyses would be approved.

In response to Khedouri's points, I said that we would do the best we could to present good analyses. However, I also pointed out the political and technological imperatives that would perhaps cause the political leadership to make a positive decision independent of the analyses we were discussing. Khedouri conceded this might be the case, but as long as there was no political direction, his job was to ask for the analytical case and to pass judgment on the programs he dealt with on that basis. I could not quarrel with his position and told him that I was grateful for giving me the opportunity to at least make the case. This meeting strengthened the conviction that I had already formed that Khedouri is an extremely valuable public servant.

On July 20, 1983, I went to Colorado Springs to attend the fourth annual launch vehicle symposium held by NASA and the Air Force. I had started these symposia when I was serving as secretary of the Air

Force in order to coordinate the plans of NASA and the Air Force for the use and the development of space launch vehicles. (The first of these symposia was held in April 1980.) The senior Air Force leadership concerned with space operations was present, General James V. Hartinger, commander of the Space Command, Lt. General Forrest S. McCartney, the head of the Space Division of the Air Force Systems Command; and from the Pentagon, there was Mr. Jimmie D. Hill, the deputy under secretary of the Air Force for Space Systems. McCartney and I were the cochairmen of the meeting, and we listened to a number of briefings both from the Air Force and from NASA about what was being done. The primary concern of the Air Force people was the development of what they called "assured access to space." What they meant by this term was that they wanted to have a fleet of launch vehicles that would allow them to launch satellites critical to the national security at any time necessary. They wanted to have a space launch system that was extremely responsive to their requirements and that was not vulnerable to failure of any of the components. The NASA delegation, of which I was the leader, could not take issue with this requirement, but we also tried to demonstrate that a shuttle fleet of four Orbiter vehicles and two launch sites would fulfill the requirement of assured access to space. (This was, of course, the conclusion we had reached in the debate with OMB over the structure of the shuttle fleet in 1978—see chapter 7.) The Air Force people felt that at the very least, a fifth Orbiter should be procured to meet their requirement and that NASA was making a mistake by substituting the space station as the major initiative instead of the fifth Orbiter.

On the face of it, this case was plausible, but the basic problem was that the Air Force could not (or would not) support a NASA request for a fifth Orbiter in very strong terms. They were afraid that if they took that step they would be asked, either by OMB or by the Congressional appropriations committees, to pay for the fifth Orbiter out of their own budget, and this was something that they did not want to do. Furthermore, NASA's proposal to build a fifth Orbiter had been only partially accepted by the OMB during the debate over the Fiscal Year 1984 budget the year before, in part because the support from the Air Force was relatively lukewarm. What had developed here was a "Catch-22" situation in which the Air Force wanted an assured launch capability, but was not quite willing to support the proposal strongly enough for fear of having to pay for the Orbiter. What the Air Force really wanted—and was somewhat more willing to pay for—was an expendable launch vehicle that would be more directly under the control of the Air Force (the Titan 34D was the preferred vehicle and a proposal to expand the capability of this vehicle would subsequently be developed). This issue is an important one that deserves further discussion, but I will defer it until the last chapter. Suffice it to say now that failure of NASA to secure permission to construct a fifth Orbiter was one reason why the Air Force would oppose the space station.

There was also another reason why the Air Force opposed the space station, and this went back to the question of whether people were really needed to conduct the space operations that were related to the national security. I have already described the problems that the Air Force had faced in initiating a manned space flight program (see chapter 7) and that the cancellation of the Air Force Manned Orbiting Laboratory (MOL) program had left scars that are still not healed. The Air Force people at the symposium flatly said that there was now "no requirement" to have people in space for any purpose related to the national security and that they, therefore, could not support NASA's space station proposal. This was an argument that we would hear with increasing frequency in the coming months, and it would eventually be made with great eloquence by Secretary Weinberger himself.

In the meantime, the process of developing an interagency position on the space station was going on—at best, by fits and starts. The fundamental opposition to the space station by the groups I have described in the foregoing paragraphs made it impossible to come to a consensus conclusion on this issue. Despite the yeoman efforts of Peggy Finarelli, all the working group could do was to develop papers that would outline the disagreements. In order to resolve this impasse, Gil Rye had formed an Interagency Group (Space), which was an ad hoc committee pegged at a bureaucratic level between the working group and the SIG (Space). The hope was that such a committee could perhaps resolve at least some of the open issues before the SIG (Space) would have to act on them. I was designated as the NASA representative on the Interagency Group (Space) and the representative from the Department of Defense would be General Richard Stilwell, a distinguished retired four-star Army officer who was serving as the deputy under secretary of defense for policy—the same post held by Admiral

Daniel J. Murphy during the Carter administration. There was a meeting of the Interagency Group (Space) on July 29, 1983, at which matters were laid out on the table. The working group had come to an impasse, and the issue was how to get something on paper that the SIG (Space) and, later on, the president, could use as the basis for making a decision on the space station. I took a very hard-line position on the matter saying that the government would look very silly if we could not even produce a paper that would lay out the pros and cons of the matter. The meeting was inconclusive, but in a telephone conversation with General Stilwell two days later, he did agree to ask his people to work out the kind of position option paper that was needed. At least we would now have the required document outlining our disagreements.

As things turned out, I had won the battle, but lost the war. On August 5, 1983, we learned that the debate had been escalated and that Secretary of Defense Weinberger had asked for a meeting with the president in order to brief him on the national security related space programs and, presumably, to argue against going ahead with the space station. I discovered later on that the activities of the Interagency Group (Space) provided only a partial reason why Secretary Weinberger had asked for the meeting. There was another meeting at the White House on August 3, 1983, which was probably much more important and which also provided a trigger for Secretary Weinberger's request. Because of the crucial part that the August 3, 1983, meeting played in shaping the president's thinking on the space station, it is important to describe the circumstances that led to the meeting in some detail.

One of the important arguments for the space station was that it would serve as a laboratory in earth orbit that could be used not only for basic scientific research, but also for the development of processes that might have potential commercial value. By August 1983 we already had some tantalizing evidence that the prospects, at least, for doing something like this were there. Ever since the fourth flight of the shuttle, a continuous flow electrophoresis experiment had been flown as a test device. The electrophoresis process had been developed in the late 1930s as a method for separating complex biologically important polymers from each other in solution by taking advantage of the fact that most of these molecules carried different electric charges and that they would therefore have different drift velocities when subjected to an electric field. When performed in a normal gravitational field, the molecules are subject to turbulent and buoyant forces caused by small temperature gradients and other distorting effects that make really efficient separation impossible. These forces are absent in zero gravity, and, the reasoning went, separation by electrophoresis should therefore be much more efficient (by three or four orders of magnitude) than it is on the ground. Accordingly, an experiment was designed to fly on the shuttle that would prove this point. The experiments have been spectacularly successful, and the improvement factors that were expected have, in fact, been realized.

What was most important from the viewpoint of ultimately developing various processes of commercial importance in space is that the continuous flow electrophoresis experiment carried along on the shuttle flights was privately funded. The money to build the equipment came from a joint venture between a large pharmaceutical house (Johnson and Johnson) and an aerospace firm (McDonnell Douglas Corporation). It was not a small investment—of the order of $50 million—and it demonstrated that even at this early stage of the game, there were things that indeed could attract the investment of significant private capital in manned space operations. (A prosperous commercial communications satellite industry has been in existence for more than two decades. The issue was not the commercial potential of space operations, but the commercial possibilities that might be inherent in *manned* space flight.)

Jim Beggs realized very early on that the commercial potential of manned space flight should be consciously explored, and he asked Phil Culbertson to look at this area as part of his other responsibilities. In 1982 Mr. Llewellyn J. (Bud) Evans, Jr., a bright young attorney with broad experience in government, was brought in to help Culbertson in this area and to eventually supervise NASA's efforts to help the commercialization of various space activities. (Evans is the son of a distinguished aerospace executive, Llewellyn J. Evans, Sr., who was for many years the chief executive officer of the Grumman Aerospace Corporation, and the stepson of Dr. Robert R. Gilruth, the first director of the Johnson Space Center.) Evans worked effectively to develop various criteria for judging when various space processes might be ripe for commercialization. Beggs also spent a good deal of his own time working on this complex and important matter, and together with Evans he was able to create considerable interest among members of the White House staff in the prospects for commercial space operations. The cabinet secretary, Craig Fuller, became particularly interested in this matter, and it was he who arranged the meeting on August 3, 1983.

My own duties as deputy administrator dealt mostly with the internal management of NASA, working on various technical issues and handling some of the relations with the Air Force because of my special experience in that area. I had only peripheral dealings with Evans and his commercialization group, so I was not really informed of what was going on. I did not know about the August 3, 1983, meeting until after it was over but immediately realized how important it was once I heard about it. As it happened, I was in Houston on August 3, 1983, to conduct a program review at the Johnson Space Center. In the evening there was a dinner party at the home of Dr. and Mrs. Milton A. Silveira. The director of the center, Mr. Gerald D. Griffin, his wife Sandy, and Max Faget and his wife were among the people who had been invited. Nancy Faget made excuses for Max and told us that he would be late because of a luncheon meeting that he was attending at the White House on the commercial potential of various space processes. When Max showed up an hour or so after dinner, we naturally bombarded him with questions. Faget told us that he had attended a luncheon meeting at the White House of some two dozen chief executive officers of large American corporations to discuss the possibilities of the commercial potential of various space operations and space processes. The president attended the luncheon and, according to Faget, had carefully listened to the discussion. It was Faget's opinion that the meeting had made an extremely positive impression on the president and that there were strong expressions of opinion among the executives of the long-term commercial potential. Jim Beggs was also at the meeting, and he confirmed Faget's opinion to me when I saw him in Washington the next day. Beggs told me that the meeting had gone just about perfectly from our viewpoint. No one made any statements regarding near-term profits so that an atmosphere of realism was established at the very beginning. However, he said, the vision for the longer-term future that the executives were describing was very exciting and that they had really engaged the president's interest. I now knew the complete reason why Secretary Weinberger wanted to have the meeting with the president.

The meeting at which the secretary of defense and his principal assistants described the national security related space programs to the president took place on August 8, 1983, in the White House Situation Room. The Situation Room is a rather small and very plain room in the West Wing of the White House, roughly at about street level of West Executive Avenue. It is a very modest room considering the momentous events that have sometimes occurred there. I cannot claim that this meeting was as important as some of the others, but for me it was critical. Secretary Weinberger is, in my opinion, the most brilliant member of President Reagan's cabinet. He is experienced, articulate, and infinitely knowledgeable. Most important, his instincts about what is important for our national defense are very sound, and he therefore enjoys the president's complete confidence. Mr. Weinberger would be a formidable opponent. The meeting of August 8 was genuinely a high-level meeting. Mr. Weinberger represented the Defense Department. Mr. William Casey, the director of central intelligence, Admiral James Watkins, the chief of naval operations, representing the Joint Chiefs of Staff, and General Williams, the director of the Defense Intelligence Agency, were all present. In addition, the senior members of the president's staff were there as well, including Judge Clark, the national security advisor, Mr. Edwin Meese III, the president's counselor, and Mr. James Baker, the White House chief of staff. Because Jim Beggs was in Europe, I represented NASA at this meeting.

Mr. Weinberger started the meeting by introducing the subject to the president. He said that it was very important to do the right thing in our space program. He mentioned the speech on defense against strategic weapons that the president had made the previous March (March 23, 1983) and said that what we would be hearing had a strong bearing on strategic defense and what could eventually be done. Finally, Mr. Weinberger said that he and his principal advisors could see no good reason for going ahead with the construction of the space station that NASA was proposing. He conceded that a space station would eventually be built because he thought that it was in the mainstream of technological progress, but he also said that we should wait. Mr. Weinberger said that there were many more important things that should be done in space before the space station is built. He then informed the meeting that the briefing would be in two parts, one by the Defense Intelligence

Agency (Lt. Col. Tom Krebs, USAF) to discuss the Russian threat and the other by the Air Force (under secretary of the Air Force, Mr. E. C. (Pete) Aldridge, Jr.) to describe our own program.

I cannot talk about the details of what transpired at this meeting because the subject matter is highly classified. However, I can say that it was a really impressive performance. I took issue with some of the things that were said, but I was clearly outgunned. I could not really hope to match the horsepower that was assembled in the room on that day. At the end of the meeting, the president asked a few questions that were perceptive and to the point. He also sensed that there was some tension in the room in view of some of the questions I had asked during the briefings. He broke the strained atmosphere in the room by saying that he never thought he would see in reality what he had seen on the Hollywood stage forty years ago on the set next to the one he was using where Buster Crabbe was playing the part of Flash Gordon. He then looked at me, winked, and said that he wanted to hear more about NASA's plans as well. I was pleased to hear that, but I wondered why he winked at me.

A few days later (August 12, 1983) there was a meeting of the SIG (Space). The purpose of this meeting was to have Judge Clark review the agency positions on the space station and then to put together some kind of a report for the president on the matter. In short, this meeting could perhaps be the culmination of the process that had been started in April. I did not go to this meeting, but Jim Beggs and Peggy Finarelli gave us a complete report. The meeting went just about as expected, given the positions we knew that the interested agencies were taking on the space station. Admiral Watkins, speaking for the Defense Department; Dr. William Schneider, Jr., under secretary of state for security assistance, science and technology; John McMahon, deputy director of the CIA; Joseph Wright, deputy director of OMB; and Philip Hughes of the vice president's staff all took positions opposed to the space station proposal. The only person at the meeting aside from Beggs who made any favorable comments was Craig Fuller, the secretary of the cabinet. The arguments at this meeting were very similar to the ones that I have already described with those representing the national security interests saying that there were no requirements for the space station and the rest arguing that the space station might be a good idea but that we could not afford it right now. We were left wondering just what Judge Clark would report to the president after this meeting, and I have to confess that I feared the worst.

On August 18, 1983, Senator John Glenn issued a formal statement of support for NASA's space station program. He had spoken favorably of the space station program before so that this was not really a surprise. However, it was welcome because we were not getting very much support from any quarter. At the time, Glenn was still a leading contender for the Democratic presidential nomination so that we hoped that this would also be a factor. We also tried to shore up our position with the White House staff. I talked with Alan Shepard, America's first astronaut, who is a friend of Mr. James Baker, the White House Chief of Staff, and asked him to see what he (Shepard) could do to help. Shepard reported back in a few days that Baker was favorably inclined but that there was a divergence of opinion among the people in the White House as to what should be done.

At the end of August we were quite pessimistic. The interagency process that was being carried out by SIG (Space) did not lead to a favorable outcome from our viewpoint. If the SIG (Space) arrived at a consensus, then it was that no space station should be built. We had enough reason to believe that the president really did want to go ahead with the project so that we kept up our campaign in spite of the fact that we really had no real support among the other agencies. At this time, also, both Jim Beggs and I had a number of conversations with senior people in the Department of Defense to see if we could not shake the negative position that they had taken on the space station. I met with Dr. Robert S. Cooper, the director of the Defense Advanced Research Projects Agency (and a former director of NASA's Goddard Space Flight Center) twice (October 3, 1983, and November 7, 1983) and spoke with him on the telephone several times to try and persuade him to support the space station. I also met with secretary of the Air Force, Verne Orr, on October 26, 1983, for the same purpose. In both cases, I failed. I was depressed by my inability to make the case well enough to persuade my friends that building a space station would be in the national interest. Beggs had been correct after all in assuming that we probably would not be able to get the support of the Department of Defense.

There were some bright spots. On September 15, 1983, the *Washing-*

ton Post carried a story by the columnist Lou Cannon that the president was likely to approve NASA's proposal for a space station. We did not know the origin of the story, but we hoped that it was leaked by someone on the White House staff who was favorably inclined toward NASA's proposal and wanted to counter some of the negative opinions being expressed at the time.

Late in September there was another opportunity for us to try and persuade the president and his staff to make an announcement that would commit the nation to the construction of a permanently manned orbiting space station. NASA had been founded in 1958; thus, 1983 was our twenty-fifth anniversary year, and some appropriate celebrations were being planned. The main event was a birthday party that was to be held at the National Air and Space Museum on October 19, 1983. We knew that the president would be there and would make a speech. On September 28 we had a meeting to plan for the event, and I even had written a draft speech for the president to make at the right moment. A week later, on October 5, 1983, Craig Fuller called to tell us that there would be no announcement on the space station at the birthday party on October 19. He hinted that there might be something in the State of the Union message, but that we should not expect anything in the immediate future.

From my viewpoint, the last weeks of September and the months of October and November were the low point in the effort to secure a commitment to the space station by the administration. Most of our efforts to secure outside support from other departments of the government had failed. As far as we knew, there was only one cabinet officer who was at all favorably disposed and that was the secretary of commerce, Mr. Malcolm Baldrige. Jim Beggs and Jack Schmitt had persuaded Baldrige that there was a long-term commercial potential inherent in the space station, and so we had at least some backing. Our efforts to persuade the people in the State Department that there was a potential in the space station program for positive international collaboration were rebuffed. And I have already dwelt upon our dismal failure to persuade anyone in the Department of Defense to support the space station proposal. Even our friends were losing heart. On September 30, 1983, Jim Fletcher came by for a visit and told us that we should think about backing away from pushing the space station project this year. He thought that the complications caused by the 1984 presidential election would make it difficult to attract the attention of the people in the White House in order for them to concentrate on something like this. I pointed out to Fletcher that he, himself, had secured a commitment to the space shuttle program from President Nixon in January 1972—also in an election year. Fletcher laughed, but my comment did not change his advice to us.

Perhaps the actual nadir of our fortunes, or at least what we thought were our fortunes at the time, was reached at our morning staff meeting on October 4, 1983. At this session, Jim Beggs wondered aloud whether Fletcher might not be right. He speculated on what might happen if we withdrew the space station proposal and substituted the proposal to commit to the construction of a fifth Orbiter instead. Both Phil Culbertson and I argued against the proposal, but we also both knew that Beggs had a point. We really were in the position of fighting an uphill battle, and there does come a time to cut losses in a situation such as this. While we considered backing away, we finally decided to stick to our guns, and two days later we were back on track.

On October 14, 1983, it was announced that Judge Clark would leave his post as national security advisor to succeed James Watt as the secretary of the interior. This was a blow because Clark had been one of our supporters in the White House. On the other hand, Clark would still be in the cabinet and might therefore still be in a good position to help. Clark was succeeded as national security advisor by retired Marine Colonel Robert C. (Bud) McFarlane. McFarlane was an excellent choice for the job, having just finished a difficult assignment as the president's chief negotiator in the Middle East. We felt slightly more optimistic that these changes were, on the whole, beneficial.

There was another positive event at this time that lifted our spirits. On October 17, 1983, Dr. Heinz Riesenhuber, the minister of science and technology of the Federal Republic of Germany, visited Washington. I had met Riesenhuber during our visit to Germany with the *Enterprise* and had talked with him briefly about the space station. (I had also had much more detailed conversations on the subject with the senior members of Riesenhuber's ministry as described in chapter 13.) These talks were now to pay off. Riesenhuber had, by now, become a strong supporter of the idea that a space station should be built as a collaborative program between the United States and Western Europe. In addition to the intrinsic value of doing the program on an international basis, there was now another more immediate reason that moved Riesenhuber: the NATO countries were about to deploy several hundred Army Pershing IRBM's and Air Force medium-range cruise missiles in Western Europe to counter the deployment of a similar number of SS-20 intermediate range missiles by the Russians in Eastern Europe. (The NATO deployment of American missiles was actually the implementation of the "two-track" negotiation plus deployment decision that had been taken during the Carter administration.) The move to emplace American missiles was quite controversial, and therefore Riesenhuber and his colleagues in the German cabinet were looking for positive things to shore up the alliance. The space station—done as a collaborative program—was a good example of what might be done.

On the evening of October 17, 1983, there was a dinner party for Riesenhuber at the Cosmos Club attended by a number of senior officials from the State Department including Under Secretary William Schneider and Assistant Secretary James Malone. Riesenhuber made a strong statement in support of the space station program, and he told the group that Germany was eager to participate in an international program with the objective of building the space station. I was delighted with this turn of events because I could see that Riesenhuber's visit had a very positive effect on the attitude of the State Department people toward the space station. Taking the *Enterprise* to Europe earlier in the year had, in fact, had the desired effect.

On October 19, 1983, there was a meeting between Jim Beggs and Fred Khedouri of OMB that was perhaps another straw in the wind. When we submitted NASA's Fiscal Year 1985 budget to OMB a month earlier, we had included a proposal to build a permanently manned earth orbiting space station in the NASA budget request and had asked for $150 million to get the project started. At the time, the response from the people at OMB was that they could not approve the item because they felt that our projected runout cost of six to eight billion dollars for the development program between 1985 and 1992 was more than was appropriate given the projected federal deficits for those years. They recognized, however, that the space station initiative would eventually be the subject of a presidential decision, so our proposal was set aside pending the decision. Khedouri was now ready to offer a compromise. He told Beggs that OMB would be willing to fund a space platform that would initially be unmanned but that might be expanded later on to include a manned capability. (This was essentially option one as defined by Jim Fletcher's committee a year earlier.) Khedouri felt that we could afford to do something like this if the total runout cost could be kept to something like two billion dollars for the development of the platform. Beggs told us that he had been noncommittal at his session with Khedouri but that he felt more optimistic now about our prospects. His opinion was that Khedouri would not make such an offer unless he thought there was a chance that NASA's proposal for a fully manned space station might succeed. That same night (September 21, 1983), the twenty-fifth anniversary celebration of the National Aeronautics and Space Act was held at the Air and Space Museum. It was a good party, and most of the people who had something to do with NASA over the years were there.

About a month later (October 19, 1983) there was a more formal celebration of the anniversary. It was held at the Air and Space Museum. The president was there and made a very nice speech. As we expected, he did not make an announcement about the space station, but he did make a strong plea for a space program with imaginative goals. We learned later that the speech had been drafted by people in Keyworth's Office of Science and Technology Policy and that Keyworth had now taken the position that NASA needed to look toward a much more expansive horizon than just a space station. As far as I was concerned that night, I was very pleased that the president himself was there, that he had good things to say about NASA, and that all of my old friends were enjoying themselves. I reflected on how far we had come since that day in October 1957 when I stood on the roof of Building 157 at the Livermore Laboratory watching *Sputnik I* cross the evening sky. I wondered whether there was ever before in human history a period of such stupendous technical development in so short a period of time. It was, of course, true that the resolution of the space station issue was still in the future, but for this evening anyway, I thought that we could well afford to bask in the light of our past achievements.

The events of the past two months had finally convinced me that we would have to try and push the space station proposal through the administration without any substantial help from anyone else in the federal government. I knew, of course, that the Apollo program was initiated by President Kennedy against the advice of many senior officials in his administration and that these circumstances were repeated when

President Nixon approved the space shuttle in 1972. I looked through some of our old files and found some of the papers that had been written by these advisors for Kennedy and Nixon. As I have said in an earlier chapter, I was particularly amused by one paper, produced by a committee headed by Professor Jerome Wiesner of MIT, that advised, among other things, that President-elect Kennedy carefully examine the Mercury program with a view toward canceling it. This paper was submitted to the president-elect on January 17, 1961, five months before Kennedy committed to the Apollo program. There was, therefore, plenty of precedent for poor advice which somehow made us feel a little bit better.

NASA has an Advisory Council that has a very distinguished membership from the scientific and engineering communities. There are also some members from the general public to provide broader advice to the NASA management on the merit of various NASA programs. As luck would have it, one of the general public members of the council was the distinguished journalist, Mr. Hugh Sidey. Sidey's specialty is the study of the presidency, and he writes a regular column on this topic for *Time* magazine. We asked Sidey to meet with us to discuss the problem we were having in persuading President Reagan to make the decision to build the space station. On October 26, 1983, Jim Beggs and I had a long lunch session with Sidey during which we were told a large number of really fascinating political stories. Among other things, Sidey told us that he thought we had a good chance to persuade President Reagan to approve the space station program. He based this opinion on his judgment that the president was looking for ways to demonstrate that the United States was, indeed, a great nation and that taking the lead in the space station program would be one way to do this. We fervently hoped that he was right. At this relatively bleak time for us, it was very heartening to have someone like Hugh Sidey tell us that we should stick to it and we would be all right. A few weeks later, Sidey would write a column that appeared in the *Time* issue of November 23, 1983, commemorating the twentieth anniversary of President Kennedy's assassination. In this piece, Sidey would tell the story of how Kennedy reached the decision to go to the moon over the objections of his closest advisors.

While the meeting with Sidey was heartening, there were others at the time that gave us real cause for concern. On October 31, 1983, one of Keyworth's assistants at OSTP, Dr. Ralph DeVries, came to see me. He sounded me out on the plans that were being made by OSTP to develop more ambitious long-range plans for NASA for the next twenty to twenty-five years. DeVries informed me that they had drafted the president's speech for the twenty-fifth anniversary and that NASA should be more mindful of what the president was saying. He also asserted that there would be no space station but that, rather, there would be a presidential commission to develop the "real" plans that NASA would eventually execute. By this time, I was honestly tired of people like young Dr. DeVries, and I hope that I was still polite when I ushered him out of my office.

Fortunately, this period of suspense would end soon enough. The Fiscal Year 1985 budget would have to be formulated before the end of the year so that some decision on the space station would have to be made soon. Accordingly, plans were now being made to hold a high-level meeting that the president would attend. We would have the opportunity to make our case for the space station during this meeting. Gil Rye and Craig Fuller came to see us on November 18, 1983, to plan for the meeting. By now Fuller was an open and strong supporter of the space station program, and he was confident that we would eventually be able to persuade the president to go ahead with it. He advised us that the proper forum for the briefing before the president would be to use the Cabinet Council on Commerce and Trade. We knew that the secretary of commerce, who chaired this council, was a supporter of the space station program. Perhaps, we thought, this would help. One of the things I did as a result of this session was to have a nice model of the space station built that we could use for the Cabinet Council meeting. The shop people at the Langley Research Center did a beautiful job in short order and, by the time of the meeting, we had a very nice scale model of the space station to show to the people who would attend.

The Cabinet Council on Commerce and Trade met to discuss NASA's proposal for the space station in the Cabinet Room on December 1, 1983. We had spent considerable effort preparing for this briefing. Norm Terrell, Peggy Finarelli, and Terry Finn were the principal people who orchestrated our arguments. The president was in the chair and a number of senior officials including the vice president, the president's counselor, Edwin Meese III, Secretaries Baldrige, Hodel, and Clark, and Attorney General Smith were there. David Stockman, the director of OMB, James Baker, the White House Chief of Staff and Dr. George A. Keyworth II were present, and the other cabinet-level officials were represented by their deputies. (A complete list of the people who were present as well as the agenda are included in appendix 10.) The model of the space station was positioned against the windows of the Cabinet Room and, as we had hoped, it attracted considerable attention. When the president walked into the room, he stopped at the model and asked me a few questions before moving on to his chair. Gil Rye introduced the NASA people (Beggs and me) and also provided a short description of the interagency process we had gone through that had culminated in this meeting. He stood on the other side of the table from where the president sat, and Jim Beggs stood next to Rye as he was being introduced. Unfortunately, my post was with the space station model, which meant that I sat directly behind the president against the windows. I therefore had a good view of Beggs during his briefing, but, unfortunately, I could not see the president's face and was therefore unable to judge his reactions on the spot.

Jim Beggs did his usual superb job in making the space station presentation. He judged correctly that this particular audience would react best to a low-key approach and that is the one he used. Beggs had no problem holding the president's attention, and he finished in about twenty minutes. The president than asked for comments from people around the table and started by requesting an opinion from his science advisor, Dr. Keyworth. Keyworth said that the space station is nothing new and that the Russians already have one. (He was referring to the Russian *Salyut*, which is a small space station that is periodically manned.) He went on to question whether we should simply catch up with the Russians by building a space station of our own. Rather, he said, we should "leapfrog" the Russians and adopt the longer-range goal of going back to the moon to establish a permanent base there. He concluded his comments by recommending that NASA's plan should be set aside until a "space summit" meeting is held to determine the right course for the nation's space program. Secretary Baldrige followed Keyworth and supported NASA's proposal. He told the group that he felt a decision should be made soon and that it should be positive because of the potential long-term economic payoff that a space station promised. The U.S. trade representative, Mr. William Brock, seconded the opinion expressed by Secretary Baldrige and strongly supported the space station proposal. This was a pleasant surprise because we did not know at the time that we had Brock's support.

The deputy secretary of defense, Mr. Paul Thayer, represented the Defense Department. He said he thought it important to conduct operations in space with people but that as yet no significant military requirements for such operations had emerged. He also expressed concern over the funds that would have to be committed to the construction of the space station because there were higher priority programs that he (Thayer) felt should be done before the space station. Thayer's money argument was echoed by Deputy Treasury Secretary McNamar.

David Stockman, the director of OMB, made an eloquent statement to the effect that he thought building a space station was an excellent idea and that it should eventually be done. He went on to argue however, that given the financial situation that the country was facing, we should defer the decision for a few years. The attorney general, Mr. William French Smith, was sitting next to Stockman. He turned to Stockman and said: "You know, David, I'll bet that the Comptroller of Ferdinand and Isabella made exactly the same speech when Columbus proposed to them that he sail west to reach the far east!" The president quickly interjected: "Yes, Bill, you're right, but you remember that the Comptroller won the argument: Isabella had to pawn her jewels in order to pay for Columbus' trip!" "What I want to know," the president went on, "is, who has the jewels around here?" There was general laughter at this quip and from then on, there was definitely a more relaxed atmosphere in the room afterward. Craig Fuller wound up the discussion by making a very strong statement in support of going ahead with the space station. He argued that we had carefully examined all the alternatives and that the process was finished. He then went on to repeat Secretary Baldrige's argument about the long-term economic gains that might result from a space station and strongly recommended that we get on with it. After Fuller's speech, the president thanked all the participants and adjourned the meeting without stating his own opinion or announcing any decision.

I have to confess that I was disappointed at the outcome of the meeting. As best I could tell, there were more people against us than for us. As we were walking out of the Cabinet Room, I told Beggs that I thought we did not lose anything at the meeting but that we also did not gain much. Beggs replied: "You're wrong. I could tell the president is with us. He winked at me a couple of times while the other people were making their critical remarks!"

XV The President Decides

Columbia was successfully launched on November 28, 1983, for the ninth shuttle mission, which was also the first flight of Spacelab. As was my custom, I was in Houston for the event. It was a letter perfect takeoff as we watched the Kennedy Space Center launch site on the television monitors from the Mission Control Center. This flight was significant because in addition to flying Spacelab, we also had the first non-American crew member onboard, Payload Specialist Dr. Ulf Merbold of Germany. I have already mentioned the Spacelab (see chapter 13) in connection with my visit to Germany earlier in the year. Now we would see it operate for the first time. I spent a day at the Johnson Space Center after the launch in order to watch the scientists working at the Payload Operations Control Center (POCC) with their colleagues onboard *Columbia*. (In addition to Merbold, Drs. Byron Lichtenberg, Owen Garriott, and Robert Parker were also on board performing scientific experiments.) I was extremely pleased at the way things were going. There was no doubt that our hopes of a close interaction between the scientists on the ground and the crew through the new Tracking and Data Relay Satellite (TDRS) communications system were more than fulfilled. There was enormous enthusiasm and elation among the people on the ground as the first results from *Columbia* came through. We could actually watch the crew members performing their experiments on the television monitors as they were being done, and the scientists on the ground could discuss the results with the crew and suggest changes in the procedures in real time.

If there was ever any doubt in my mind about the value of what could be done by taking advantage of the presence of human intelligence and judgment in space, they were resolved by watching these people at work. What was more important is that I began to recognize that what I was watching here was a foretaste of what we could do once we had a space station. It is surprising that I had not realized earlier that Spacelab and its operation would be a natural precursor to the space station and that we should consciously begin to look at the operation in that way. The Spacelab module was designed to fit in the payload bay of the shuttle, and this too would be important for using it as a building block for the space station since anything that could not be easily transported to the orbital position of the space station using the shuttle was not likely to be very useful. The notion, therefore, that a suitably modified Spacelab module such as the one in which Merbold, Garriott, Lichtenberg, and Parker were now doing their experiments should be used as a building block for the space station was driven home in the most graphic way by what I saw. And the fact that Spacelab was a European product argued for trying to bring the Europeans and other friends and allies abroad into the space station program.

At this time, no decision with respect to the space station had yet been made nor was there a consensus on whether the program should be done on an international basis. Beggs and I had laid the foundation during our visits to Europe in May and June to open that option, but we were obviously in no position to make any commitments. As it happened, by late November, some people in the administration were looking for something favorable to do together with the Europeans to counter the unfavorable publicity caused by the deployment of American Pershing and cruise missiles (see chapter 14). Beggs had always been anxious to have the president participate in the flights of the shuttle ever since the success of the president's visit to the Johnson Space Center during the second mission (see chapter 9). This time, also, because we were flying the first non-American in the crew, preparations had been made for the president to talk to the crew while they were in orbit. But, there was an extra attraction: we could arrange things so that not only the president but also Chancellor Helmut Kohl of the Federal Republic of Germany could talk to the crew at the same time. Thus, the leaders of two of the important members of the Western alliance would be linked together through Columbia's communications loop and would talk to each other as well as to the crew—which had both German and American members. This linkup was successfully accomplished on December 5, 1983, and it had the desired effect in that the event gave rise to much favorable publicity for the United States in Germany and elsewhere in Europe as well. It would also influence the president's thinking on the formulation of the space station program.

The day after the meeting of the Cabinet Council on Commerce and Trade, there was a meeting of the White House Program Resources Board. This was the high-level review board for budgetary matters with James A. Baker III, the White House Chief of Staff sitting as chairman, and Counselor Edwin Meese III and OMB Director David A. Stockman as members. The purpose of this meeting on December 2, 1983, was to review the NASA budget. Naturally, we were most anxious to learn what decision, if any, had been made with respect to the space station, but we were to be disappointed. After Jim Beggs made his usual lucid presentation of the NASA program to the board, we were told that we would get essentially what we had requested in our Fiscal Year 1985 request, but that the $150 million item for the space station would require a special meeting with the president before a final decision could be made. We felt very good about the overall budget and prepared to make our case for the space station with the president as best we could. After the Program Resources Board meeting, Gil Rye told us that he thought the president was very favorably disposed to making a positive decision. Maybe the many months of work we had devoted to this outcome would finally yield results.

The next day was a Saturday (December 3, 1983), and I had to fly to California for a speech before the San Francisco Chapter of the Air Force Association that night. When I arrived there, I discovered a message to call Jim Beggs. When I reached him, he told me that he had a follow-up meeting with Fred Khedouri of OMB to nail down the details of the Fiscal Year 1985 NASA budget. Khedouri had told Beggs that the $150 million we had requested would be in the president's budget and that this would be confirmed the following Monday (December 5, 1983) at a decision meeting with the president. So, we had finally won the debate and the decisive vote in the administration —the president's—was in favor of going ahead with the commitment to build a space station. Needless to say, I was elated by our victory. I also reflected as I was standing at the pay telephone in the San Francisco airport talking with Beggs that this was a historic occasion. I told

Beggs that only twice before, in May 1961 for Apollo and in January 1972 for the space shuttle, had a president set a major new direction for the American space program. This new initiative meant, among other things, that Jim Beggs would be remembered along with Jim Webb and Jim Fletcher as one of the NASA administrators who had succeeded in persuading a president to do something brand new. It also spoke well for the political vision and sense of history that President Reagan possesses. Just as his predecessors had done, he had taken the decision to go ahead with the space station against the advice of most of his principal assistants.

I made my speech in San Francisco that night and then went to Los Angeles to attend a meeting of space enthusiasts that had been organized by the prominent author, Dr. Jerry Pournelle. It was a very broadly based group that included *Apollo 11* astronaut Edwin (Buzz) Aldrin, Dr. Lowell Wood, a senior physicist at the Lawrence Livermore National Laboratory, the author and engineer Mr. G. Harry Stine, Dr. David Criswell of the University of California, and many other people prominent in the space industry. This group meets once or twice a year and produces a report on what the members believe should be done in the space program. I gave the group an optimistic report on where we stood with the space station and told them that I thought a decision would be forthcoming from the president in a couple of days. I was rather surprised at the reaction of the group. There was a general feeling that the space station was a good thing but that it was not ambitious enough as a goal. Many people present felt that we should make the commitment to go back to the moon now and to establish a permanent base there. There was a spirited debate on this issue, and I was generally on the losing side. I tried to explain that, in our judgment, the political system in Washington was simply not ready for a bigger program, but I could not convince most of the people present. I did finally succeed in persuading the group to put a statement supporting the space station in their report, but I could not prevent some statements critical of NASA for not being farsighted enough from also being included. This was, of course, the same criticism that Dr. George Keyworth, the president's science advisor, was now making as well. I mention this meeting only to point out that we had opposition not only from those who thought we were trying to do too much, but also from those who believed that we were too timid.

Jim Beggs had his budget meeting with President Reagan on Monday, December 5, 1983. I was not present at the meeting, so I cannot discuss the details, but it was at this session that the final and formal decision was made to include the space station in the president's Fiscal Year 1985 program. The question of whether the space station should be an international effort was left open for future action. Finally, it was decided that the president would make some kind of a public announcement of the commitment to building the space station soon, possibly as part of the State of the Union message that the president would deliver in January.

For the remaining weeks of 1983, I spent most of my time on internal NASA planning deciding how we would execute the space station program. We had to develop an organization both at NASA headquarters and at the various NASA centers that would be capable of managing the program we had in mind. In several discussions with Phil Culbertson and Jim Beggs during this period, we concluded that we should establish a Program Office in Washington to be headed by an associate administrator to handle the relationships with OMB, the other federal agencies, and the Congress. The Washington Space Station Program Office would also have the overall responsibility for formulating the general technical concepts. The execution and project management would be delegated to the Johnson Space Center in Houston, Texas. This organizational pattern was similar to the one that had been successfully applied in the space shuttle program, and we felt that it would also work effectively for the space station. We decided that we would go ahead and establish the "lead center" for the space station at the Johnson Space Center as soon as we could reasonably expect to set up the organization. We were anxious to involve all of the NASA centers in the effort, and so it would take some time to develop the appropriate organization, but it was very important to get started. We also decided not to establish the Washington Program Office until the Congress had a chance to act on the President's Fiscal Year 1985 budget. We did not want to look as if we were preempting the congressional prerogative of acting on the president's proposal.

During these weeks, we also had a number of conversations with some of the people in the science advisor's office to see whether we could find some kind of a common approach to the space station. Dr.

Richard Johnson, a distinguished space physicist who had spent many years at the Lockheed Corporation in Palo Alto, had joined the science advisor's staff. I had known Johnson during the years I worked at NASA's Ames Research Center, and I felt pleased to have a person in the science advisor's office who had a thorough and sophisticated under-standing of NASA and the space program. I had several meetings with Johnson during this period to explain to him what we had in mind with respect to the space station program. During one of these meetings (December 30, 1983), Johnson told me that the president would not mention the space station in the State of the Union message as we were hoping. I reported this to Beggs, and we redoubled our efforts to work with other members of the White House staff to get some kind of a statement included in the president's speech. On January 4, 1984, Beggs and I met with Johnson and the science advisor, Dr. Keyworth, to discuss the situation. The meeting was inconclusive. Keyworth still wanted to have a more ambitious goal for the nation's space program than the space station we were proposing. He told us that he thought the establishment of a permanent base on the moon would be a more appropriate goal, and he argued that we should delay the announce-ment of the space station until such a time at which the moon base proposal would be ready for presentation. He said that he was ready to support the space station proposal but that he wanted to do this as part of a more comprehensive plan.

I was not involved in the details of preparing the president's State of the Union message, nor did I participate in the debate on whether to make the space station an international effort. Sometime in January the president's senior advisors decided to include the space station as one of the major proposals the president would make in the State of the Union message. We also learned that the president would invite our friends and allies around the world to participate with us in the develop-ment and construction of the space station. I very badly wanted to hear the president make the announcement of the space station in his State of the Union speech, but it turned out to be very difficult to get tickets to attend. Up until the very last minute, I did not know whether I would be able to go. Fortunately, the chairman of the House Science and Technology Committee, Representative Don Fuqua, (D., Florida), who is a good friend, and one of his senior staff members, Mr. Thomas

Tate, were able to secure a ticket for me shortly before the president was scheduled to make his speech. I was extremely grateful for this because I wanted to be there to see for myself the culmination of the efforts we had made during the past three years.

The State of the Union speech by the president before the Congress is one of the festive occasions in Washington. Actually, the delivery of such a message is required by the Constitution which says in Article II, Section 3, that the president "shall from time to time give to the Congress information of the State of the Union and recommend to their consideration such measures as he shall judge necessary and expedient." Until the presidency of Franklin Roosevelt, these messages were delivered in writing. It was he who started the tradition that the president himself would go to the Capitol building and deliver a message early each year on the State of the Union in person before a joint session of Congress. The speech was scheduled for 8:00 P.M. on Wednesday, January 25, 1984. I walked up to the Capitol Building from my office, which was in a building just a few blocks away. The Chamber of the House of Representatives was filled to capacity. The representatives and senators were all there as well as the cabinet, the Supreme Court, the military leaders, and the members of the diplomatic corps. The entire leadership of the nation was assembled to hear the president. I sat in the gallery—which was packed with visitors and members of the press—and waited for the president to arrive. There is a formal ritual that is played out when the president visits the Congress —the coequal branch of the government. An escorting committee of senators and representatives is announced by the Speaker of the House. Then, at the appropriate moment, the doorkeeper of the House announces in a loud voice, "Mr. Speaker, the President of the United States," at which point the president and the escorting committee walk down the center aisle and the president steps to the rostrum. Then the president starts to speak.

President Reagan is a forceful and articulate orator, and I could tell after a minute or two that this speech would be similar to the inauguration speech that I had watched on television in my Pentagon office almost exactly three years earlier. Once again, he emphasized the fact that we are a great nation and that we must act like one. After a brief outline of recent history, he delivered what was the most memorable line in the speech: "America is too great for small dreams." He then went on to outline the four goals he had set for what he called "freedom's next step:"

1. We can ensure steady economic growth.
2. We can develop America's next frontier.
3. We can strengthen our traditional values.
4. And we can build a meaningful peace.

I was elated when I heard the president announce this list of priorities. Not only would the space station be mentioned in the State of the Union message, but if my guess about his second point turned out to be correct, the space station would become one of the four major elements of the speech. I waited for the passage about the "next frontier," and when it came, I was not disappointed. Here is what the president said:

> Our second great goal is to build on America's pioneer spirit and develop our next frontier. A sparking economy spurs initiative and ingenuity to create sunrise industries and make the older ones more competitive.
>
> Nowhere is this more important than our next frontier: space. Nowhere do we so effectively demonstrate our technological leadership and ability to make life better on earth.
>
> The space age is barely a quarter of a century old, but already we've pushed civilization forward with our advances in science and technology. Opportunities and jobs will multiply as we cross new thresholds of knowledge and reach deeper into the unknown.
>
> Our progress in space, taking giant steps for all mankind, is a tribute to American teamwork and excellence. Our finest minds in government, industry and academia have all pulled together, and we can be proud to say: we are first, we are the best and we are so because we are free.
>
> America has always been greatest when we dared to be great. We can reach for greatness again.
>
> We can follow our dreams to distant stars, living and working in space for peaceful, economic and scientific gain. Tonight, I am directing NASA to develop a permanently manned space station and to do it within a decade.

A space station will permit quantum leaps in our research in science, communications and in metals and life-saving medicines which can be manufactured only in space.

We want our friends to help us meet these challenges and share in their benefits.

NASA will invite other countries to participate so we can strengthen peace, build prosperity and expand freedom for all who share our goals.

Just as the oceans opened up a new world for clipper ships and Yankee traders, space holds enormous potential for commerce today.

The market for space transportation could surpass our capacity to develop it. Companies interested in putting payloads into space must have ready access to private sector launch services.

The Department of Transportation will help an expendable launch service industry to get off the ground. We will soon implement a number of executive initiatives, develop proposals to ease regulatory constraints and, with NASA's help, promote private sector investment in space.

There it was: the chief executive was making a commitment that we had been waiting for—many of us for a long time. We would build a space station, we would do it within the decade, and we would do it with the help of our friends and allies abroad. I sat there in a reverie remembering when I had first read about space stations more than forty years ago. Now I might even live to see this dream become a reality.

After he finished his speech, the president received a standing ovation. There was a warm glow in the room because it was a positive and challenging message. The people in the House Chamber felt ready to respond. As I made my way out into the jammed hallway, I ran into Mrs. Joan Clark, who had been my partner on the television shows in Vienna eighteen months before. "Well," she said, "you finally got your space station!" All I could do in reply was to grin—and I hope that I did not look too foolish.

XVI The Congressional Debate

In the fifteenth century, Thomas à Kempis wrote, "Man proposes but God disposes." Someone translated this to Washington politics by saying that "The President proposes but Congress disposes." It may be a sacrilege to compare Congress to God, but as far as we were concerned, we could only agree that it was accurate. The president had proposed the space station program, and now it was up to the Congress to "dispose" of the proposal and to make the final judgment on what would be done. The six months following the president's State of the Union message were devoted almost exclusively to pushing the space station proposal and the rest of the NASA program through the Congress.

There is now strong and widespread support for a vigorous space program in the Congress. This has not always been the case and, especially in the late 1960s and early 1970s when there was a general feeling that we were spending too much on space, budget restrictions really hampered our progress. This had changed by the early 1980s, and the support was now there. This did not mean that the Congress would accept in an uncritical way what NASA proposed. Quite the contrary, along with stronger support came stronger interest, and with interest came a more critical and knowledgeable analysis of the proposed NASA program.

In January 1984 we knew that there would be considerable opposition to the president's space station proposal. There were three major arguments against the space station on which the opposition centered. One argument was that NASA's space station program was premature and that a fifth shuttle Orbiter should be built before a commitment to a space station was made. The feeling among people who made this argument was that a fifth Orbiter was needed to give the shuttle fleet sufficient capability to survive the loss of a shuttle vehicle and still be able to execute the missions that were being planned at the time. Also, some people felt that the full on-orbit operational flight profile of the Orbiter would not be fully explored unless the vehicle could spend more than a week or so in earth orbit. The people who believed this therefore supported a proposal to extend the time duration on orbit for the vehicle from a week to two months. This proposal was called the "Extended Duration Orbiter," and it was favored especially by a number of members of the House Science and Technology Committee. Finally, there was the opinion that by giving NASA the space station now, the attention of many good people in NASA would be diverted from the work of making the space shuttle fully operational.

The second set of arguments had to do with the value of putting people in space. This same question, which had already been thoroughly debated within the administration, would now be revisited once again, sometimes with a quite different emphasis than in the earlier discussions. The man-in-space question would be particularly important when the administration's space station proposals came before the appropriations committees of the House and the Senate. In fact, the subcommittee of the Senate appropriations committee that handles NASA's program (the Subcommittee on Housing and Urban Development-Independent Agencies) put the following statement in the report that accompanied the NASA Fiscal Year 1984 Senate Appropriation Bill:

> In reference to the Space Station, the Committee suggests that NASA devote additional effort to exploring the potential benefits that can be derived through the design of a fully automated space platform. The Committee believes that an evolutionary approach to a manned Space Station is the most effective way of proceeding. In addition, the Committee expects that the development of an automated platform might lead to significant advancements in pattern recognition, robotics, and artificial intelligence.

This argument was the same one that Keyworth had made two years earlier, before he decided that NASA was not doing enough. In any event, we had been put clearly on notice that there would be opposition to a program to construct a permanently manned orbiting space station.

Finally, there was a third contention made by many people to the effect that the commitment to a space station would "take money away" from other things that NASA was supposed to be doing. This argument is related to a more general one that the "big projects" done by NASA are done at the expense of smaller but more valuable scientific programs. Although it is widely believed, this assertion is not supported by the historical trends of the NASA budget. Generally speaking, the scientific part of the NASA budget is large when the total NASA budget is large, and it is small when total NASA spending is small. The largest amount NASA ever spent on scientific research occurred in 1967, which was the peak spending year for the Apollo program. (The 1967 budget was the largest ever for NASA, roughly $18 billion in 1984 dollars compared to the $7.5 billion NASA is spending this year.) The fact is that the NASA budget (or the federal budget, for that matter) is not really allocated according to the rules of a zero-sum game so that the trade-offs between major program elements that people allude to really never occur. Nevertheless, we would hear the statement that the "space station will destroy science" over and over again.

In addition to the three arguments I have just outlined, there was one other more general concern that would cause us problems during this period. This was the issue known as "the militarization of space." The Space Act of 1958 establishes NASA specifically as a civilian agency but also requires information sharing between NASA and the military agencies responsible for aeronautics and space. Various administrations have interpreted these provisions in different ways, and in the Reagan administration the watchword was contained in the July 4, 1982, space policy. The civil and military space programs were to be "independent but closely related." Many members of the Congress—mostly, but by no means all from the Democratic side of the two houses—were concerned about the spread of weapons in space. There had been efforts to write agreements with the Russians on "banning" antisatellite weapons (see chapter 8), and there were people who felt that some kind of negotiation with the Russians should be resumed on this subject. Furthermore, President Reagan's speech of March 23, 1983, in which he proposed that we initiate a serious research and development program to see if a defense against strategic missiles can be built, triggered even greater concerns. In spite of the fact that there are good technical reasons to believe that a defense against ballistic missiles can eventually be built, there are many people who believe that it should not be done. People in the Congress—both among the members and their staffs —who were concerned about this matter questioned us about the space station and its relationship to the development of a defense against strategic weapons. They feared that the space station would be used by the military for that purpose, and some even said that NASA was being used as a "cover" by the military to build a space station that would then be turned over to them. Considering what I have said about the attitude of the military establishment toward the space station, these questions seemed bizarre to us, to say the least. They also illustrate one of the most destructive trends in modern American politics: the deep suspicion of the military in certain quarters of the political establishment.

Those people who were not completely blinded by antimilitary prejudice knew better. The fact that influential civilian and military leaders had opposed NASA's plan for the space station was well known. Ironically, this would now be of help to us in the Congress, especially in the House of Representatives where there was considerable feeling against the military. There is a very real question in my mind whether we could ever have pushed the space station through the Congress if we had strong support from the military. As things turned out, it was probably just as well that I failed in my attempts to persuade Robert Cooper and Verne Orr to help us.

The opposition that we expected quickly became apparent during the first hearings on the Fiscal Year 1985 budget that followed shortly after the president's speech. The Subcommittee on Space Science and Applications of the House Science and Technology Committee held its hearings on February 1, 1984. Representative Harold Volkmer (D., Missouri) was in the chair. All of the questions I have outlined were aired at the hearing with particular emphasis being placed on the question of the fifth Orbiter. Several members of this committee were particularly concerned about this problem and asked some penetrating questions about this matter. Hearings of this kind are never decision meetings, so there was no outcome, negative or positive. However, it was clear to us that it would be most important for us to start making detailed contacts with individual members of this committee and with others that had jurisdiction over our programs in order to explain to them what we had in mind. Hearings were important, but on a matter such as the space station, personal contact was essential.

The next hearing was held before the Senate Budget Committee. Although the Budget Committee is not part of the formal authorization and appropriation process, it is supposed to oversee the entire Congressional budgeting operation. Since the hearing on February 27, 1984 of the Senate Budget Committee was concerned not only with NASA but with other budget items in basic scientific research, there were several other witnesses. The chairman, Senator Pete V. Domenici (R., New Mexico) opened the hearing with a statement on the importance of basic research and the development of new technology for the economy. The president's science advisor, Dr. George A. Keyworth II, was the first witness. Keyworth read a short statement in which he advocated a strong basic research program and strongly supported the administration's position on the space station. Keyworth is a "good soldier" and supported the president's program wholeheartedly even though the decision on the space station had gone against him in the debate within the administration. Jim Beggs was up after Keyworth and presented the NASA program. There were some questions afterward along the same lines we had heard from the House Science and Technology Committee.

The most interesting aspect of the Senate Budget Committee hearing was the testimony we heard from two nonadministration witnesses, Dr. William D. Carey, the executive officer of the American Association for the Advancement of Science, and Dr. John H. Gibbons of the congressional Office of Technology Assessment (OTA). Both, at least by implication, opposed the space station proposal. Carey asserted that large defense and NASA programs would "eat up" scientific research funding and that, therefore, the large programs should be stopped or not started in the first place. No one caught the nonsequitor in Carey's logic or even tried to dispute his point. Gibbons presented a highly structured argument in which he listed and elaborated all of the important points against going ahead with the space station. He started by asserting that little of scientific value would come from the space station, he criticized NASA's management approach, and said that not all of the possible options had been examined by NASA, particularly the use of automated systems. At the end of his speech, Gibbons said that the OTA could not yet take a position on the space station because there had not been enough study of the matter.

Chairman Domenici was not satisfied with this conclusion and pressed Gibbons for a "yes or no" answer on the space station. Gibbons steadfastly refused to commit himself one way or the other until, finally, an exasperated Domenici wondered out loud what OTA was supposed to be doing and what it was good for. Senator Lawton Chiles (D., Florida) picked up on this and asked Gibbons whether he thought that we would have ever gone to the moon if we had applied the kind of analysis that Gibbons had provided to the committee. Gibbons said he saw no reason why not, but it was clear from his reply that he was completely oblivious to the irony in Chiles' question. As I was walking out of the room with Phil Culbertson, I could not help but remark on the large amount of wooden-headed thinking that passes for analysis in Washington.

The next day (February 28, 1984), we had our hearings before NASA's Senate Authorization Subcommittee. (This is the Subcommittee on Science, Technology and Space of the Committee on Commerce, Science and Transportation.) Of all our congressional oversight committees, this one was most friendly toward the proposed NASA program. Senator Slade Gorton (R., Washington), who heads this committee, led the discussion. The hearing was very friendly and several members, including Gorton, Senator Paul Trible (R., Virginia), and Senator Howell Heflin (D., Alabama) made statements in support of the space station. We knew then that we would have no problem with this group.

The authorization hearings were now finished, and we spent the next month in intensive discussions with members of the House Science and Technology Committee and also with some of the Senators on our authorization committee. During these meetings, we made it as clear as we could what the space station was all about. It was during this period that NASA's Office of Congressional Affairs proved its worth. Simply scheduling the meetings with the various members of the Congress for Jim Beggs and me was a difficult chore. They also, however, helped to develop the strategy and tactics we would use to influence the individual members on the issues that were important to us. Mr. John F. Murphy, a former member of Senator Barry Goldwater's staff and wise in the ways of the Congress, headed the office and oversaw the overall effort. His deputy, Mrs. Nadia McConnell, a bright and articulate transfer to NASA from the Federal Emergency Management Agency (FEMA), would handle the day-to-day operations and Ms. Mary D. Kerwin would be her assistant. Although not a member of the Congressional Affairs Office himself, Dr. Terry Finn would also play an important part in the process to persuade the Congress to accept the space station program. Finn had headed the NASA Congressional Affairs Office in the Carter administration and now had an important position on John Hodge's space station task force. Finn was brought in specifically to help us make contact with senior Democratic members of the House of Representatives, and he turned out to be most effective in that role.

Some of the discussions we had with members of the Congress during March were particularly interesting because of the political issues that the space station proposal had stirred to the surface. There was, for example, the liberal Democrat, Representative George Brown from Southern California. Brown is a large, disorganized-looking man with a sharp mind and a well-defined viewpoint. He is a staunch supporter of technological development and the space program. During our meeting (on March 8, 1984), his major concern was not any technical or funding issue, but whether we thought that we could get the Russians to work with us on the space station. I told Brown about the collaborative efforts that were in place with the Russians on data exchange in our manned space flight programs, but I also explained why I did not think doing the space station together with the Russians was a particularly good idea. We finally got Brown's support. There were meetings with people like Representative Mike Andrews (D., Texas) whose district adjoins the Johnson Space Center in Houston and who was most helpful during the entire process. Our efforts paid off when, on March 13, 1984, the first of our authorization subcommittees, the House Subcommittee on Space, Science and Applications, approved the space station proposal with the full funding ($150 million) that we had requested. The full House Science and Technology Committee followed with its approval seven days later on March 20, 1984.

At about the same time—early March 1984—Jim Beggs went to Europe for two weeks to begin the process of developing the international aspect of the space station program. This would become important later on in the congressional debate because the international collaboration was one of the very attractive features of the space station program for many members of the Congress. The subject of international collaboration—as well as many others that affected our legislative strategy—was discussed in detail late in the afternoon of March 14, 1984, at a meeting held in Craig Fuller's White House office. The NASA people as well as those from White House Legislative Affairs were present, and so a thorough and knowledgeable discussion followed. The essential point we tried to drive home to the White House contingent was that our real congressional problem was in the appropriations and not in the authorization committees. Craig Fuller once again proved to be a strong friend, and he made a strong commitment to us to help in whatever way we felt would be most useful.

If there was any doubt in my mind that the appropriations process would indeed be the difficult hurdle, it was dispelled during a lunch meeting I had with Dr. Wallace G. Berger, the chief staff man on Senator Garn's NASA Appropriations Subcommittee, on March 20, 1984. (The full name of this Subcommittee is the Subcommittee on Housing and Urban Development–Independent Agencies.) Berger reminded me of the statement in the Senate Appropriations Committee Fiscal Year 1984 Report in which the committee suggested that NASA build an unmanned platform rather than a manned station. He wanted to know why we did not pay more attention to the committee's suggestion. It was his intention, he informed me, to recommend to Chairman Garn that the committee reject the NASA proposal for a space station. Berger is an extremely intelligent person, and I was very much afraid that his opposition would prove to be a formidable obstacle. Berger's argument that an automated space station would have more "spin-off" to the American robotics industry was especially difficult to answer because it had a certain plausibility. I tried to explain to Berger that every spacecraft that NASA builds, either manned or unmanned, is very highly automated—and I cited the space shuttle as an example. I told Berger that I did not think one could make a division that could define whether the manned or the unmanned part of the American space program had more of an impact on the robotics industry. Our discussion of this point was inconclusive. When I made the political and international arguments in favor of going ahead with the space station program, Berger simply refused to accept their validity. He also would not accept that the value of putting people in space to perform the various functions we had in mind had been demonstrated. I had drawn a complete blank.

A couple of weeks after my disastrous lunch meeting with Berger, *Challenger* was launched to perform a mission that I hoped would help us in our final effort to convince the Congress to approve the space station. I mentioned in chapter 11 the plan that we had developed in 1981 to pick up and repair a satellite that had failed on orbit. This satellite, the Solar Maximum Mission (SMM) satellite or "Solar Max," was tumbling slowly out of control in its orbit because of a failure in its electrical power system. We thought that going up to retrieve and repair it would be the clearest possible demonstration of the intrinsic value of putting people in space. Now, almost three years after we had conceived the mission, we were ready to go. The timing could not have been better because we were once again being challenged on this critical point. The attempt to retrieve the Solar Max was scheduled for April 8, 1984, and Dr. George D. "Pinky" Nelson was to be the one to go out and stabilize the satellite and maneuver the satellite into *Challenger's* payload bay where it would be repaired. This attempt was a failure, and all of us were bitterly disappointed. It turned out that the grappling fixture had not operated properly so that Nelson could not attach himself to "Solar Max" and execute the necessary maneuver.

There now followed some feverish planning to see if there might be another way to rescue Solar Max that would not involve the maneuver that Nelson had attempted earlier. By a really remarkable intellectual tour de force, the people who had built the satellite at the Goddard Space Flight Center in Greenbelt, Maryland, developed an alternate technique that called for direct grappling of Solar Max using *Challenger's* Remote Manipulator Arm. On April 10, 1984, Mission Commander Bob Crippen and Mission Specialist Terry Hart, who operated the arm, succeeded in retrieving Solar Max and placing it properly in *Challenger's* payload bay. The next day, Solar Max was successfully repaired by mission specialists George D. Nelson and James D. Van Hoften and redeployed into its proper orbit. The Solar Max mission had been a triumphant success. We had not only demonstrated that we could retrieve and repair a satellite on orbit, but we also showed that we had the flexibility to recover from a failed retrieval attempt. This flexibility, of course, is precisely what we meant when we talked about the importance of having people—and their intelligence, judgment, and imagination—in space operations. While I did not believe that our successful execution of this mission would persuade diehards like Berger,

I hoped that others would look at what we had done and draw different conclusions.

While we were trying to deal with Wally Berger and the Senate Appropriations Subcommittee, we also had to start the effort to persuade the NASA House Appropriations Subcommittee to approve the NASA program. This Subcommittee on HUD–Independent Agencies is headed by one of the most senior and distinguished members of the House of Representatives, Representative Edward P. Boland (D., Massachusetts). Mr. Boland was one of the early and strong supporters of the space shuttle program in the Congress, and I doubt whether the Shuttle program could have been executed as it was without his consistent support. Mr. Boland's chief staff man on this subcommittee is Mr. Richard N. Malow. Malow is an extremely intelligent and knowledgeable person who has also been a very strong supporter of the space shuttle program over the years.

On March 27 and 28, 1984, we had our hearings before Mr. Boland's subcommittee. It was a very detailed and lengthy session at which the entire NASA program was discussed in great detail. All of the questions about the space station that I have already discussed were aired. The manned versus unmanned space operations issue was the subject of a lengthy discussion as was the matter of possible military use of the space station. All of the members of the subcommittee were present at the hearing so that we could get a good feeling for who was interested in the various aspects of the program. It was difficult to tell from the hearing hcw the subcommittee would finally act on our proposals, but the detailed and knowledgeable questions from the members raised my hopes. One good omen was that on March 28, 1984, the House of Representatives passed the bill containing the NASA authorization by a margin of 389 to 11. The first hurdle had been passed, and we hoped that this would add momentum to the remainder of the process.

The hearings before the Senate Appropriations Subcommittee were held on March 29, 1984. In contrast to the hearings before the House Appropriations Subcommittee, the Senate hearings were perfunctory. The subject of the space station was not even explicitly discussed, although there was an implicit reference to it during a discussion of the report by the Office of Technology Assessment. We knew, of course, that the chief staff man of the subcommittee, Wally Berger, had strong reservations about the space station, but we did not yet know just how he would influence the chairman, Senator Garn, and the other members of the subcommittee.

On April 25, 1984, I had a long meeting with Malow and discovered to my dismay that there was strong opposition to the space station program on the House subcommittee as well. Chairman Boland himself was not convinced that the space station proposal was a good idea, Malow told me, and he also said that the senior Republican member of the subcommittee, Representative Bill Green of New York, was strongly opposed to the NASA proposal. In addition, Malow said several other members also had some severe reservations. I judged, after listening to Malow, that our problem with the House Appropriations Subcommittee would be more severe than the problem we had in the Senate, which was already bad enough. The Senate was controlled by the Republicans, and so Chairman Garn, we felt, would ultimately respond to a request from the White House to support the space station proposal. This stratagem was not available to us in the case of the House. What we had to do in this case was to visit every member of Mr. Boland's subcommittee and to explain to each of them in detail what it was that we had in mind. In this effort, Representative Jerry Lewis (R., California), who was a strong supporter of the space station program was most helpful. Terry Finn was also very important because he used his wide contacts among the Democrats on the subcommittee and the senior members of their staffs. During the next two or three weeks, Jim Beggs and I, between us, met at least once with each member of the subcommittee to explain to them the details of the NASA program. One of the arguments that we used to good effect was the fact that the space station program would be done on an international basis.

As luck would have it, one of the periodic economic Summit Meetings of the Western European allies was scheduled to be held in London in June. Craig Fuller, Gil Rye, and some of our other friends on the White House staff conceived the idea of using the space station initiative as one of the things that would be discussed at the Summit Conference. The president would try to get formal commitments from the Allies to participate in the program at the London meeting and that the space station initiative would be included in any communiqué that would be issued from the Summit Meeting. Once all of this was in the plans, we were able to make a stronger and more immediate case about the international value of the space station and that it was indeed a positive step in the conduct of our foreign policy.

While we were conducting the talks with the members of the House appropriations committees, we also received considerable help from the Congressional Space Caucus. This was a loose coalition of over two hundred members who were interested in a vigorous American space program. The caucus was bipartisan and was chaired by Representative Daniel Akaka (D., Hawaii). Representative Newt Gingrich (R., Georgia) served as vice chairman. Members of the caucus were most helpful in advising us on how best to frame our arguments. Ms. Diana Hoyt, the staff chief of the caucus was particularly active in this respect, as was Ms. Joyce Freiwald on the staff of Representative Manuel Lujan, Jr. (R., New Mexico). By the beginning of May, we felt fairly certain that we had four of the nine votes on the appropriations subcommittee and that we had a good chance of getting the fifth. This was the vote of Representative Louis Stokes (D., Ohio) whose district is close to the NASA Lewis Research Center. We made a concerted effort to persuade Mr. Stokes that the space station program was indeed a good thing to do. We succeeded just before the scheduled meeting of the subcommittee to consider our appropriations bill.

On May 9, 1984, the Senate passed the bill containing NASA's authorization essentially the way we had proposed it. There were minor differences between the House and the Senate versions of the bill that would have to be resolved, but both of the bills included the space station provision with the $150 million we had requested. We would thus be authorized to go ahead with the program, but the funds still had to be appropriated. The House Appropriations Subcommittee met on May 14, 1984, to consider the NASA appropriation. We knew that we had the votes now to overcome the opposition of Chairman Boland and the senior Republican member, Mr. Green. As it turned out, the matter never came to a vote because both Mr. Boland and Mr. Green also knew the vote count. The subcommittee approved our request for $150 million for the space station and put in a provision that $15 million would have to be spent on studies to see whether it would not be better to start the space station with an unmanned platform that would then evolve to a permanently manned space station later on. We all felt that we could live with this provision and felt that we had won a major victory by securing the support of Mr. Boland's subcommittee. We now still had to get the Senate Appropriations Subcommittee to agree.

On May 21, 1984, I met with Ms. Carolyn Fuller, who is the chief aide of Senator Walter Huddleston (D., Kentucky). Carolyn Fuller is an old friend who had helped me with the problem of getting the funds to construct the west coast launch site for the shuttle in the Air Force budget four years earlier. Her advice now was to follow the strategy that Jim Beggs had already outlined. We would once again try to meet with all of the members of the subcommittee, and we would also see if we could get the people in the White House to influence the Republican members in favor of approving the president's request for a space station. On May 29, 1984, we learned that the Senate Appropriations Subcommittee had acted on our bill, that they had approved the funding for the space station, but that they had also added some restrictive language that would require us to spend a fraction of the money on automated technology that would be applied to the space station. We had thus also won the $150 million provision for the space station in the Senate Appropriations Subcommittee but with a somewhat different provision in the bill for how some of the money would have to be spent than was the case in the House bill. The difference between the two would have to be resolved in conference between the two houses of Congress.

At the time, we felt rather frustrated by these complications, but looking back on it now, we had really won the debate and the remaining issues that had to be resolved were minor ones compared to the principal objective. We had won the point that the space station, roughly as we had conceived it, would be approved and that was the important thing. The only remaining problem was to agree on the statement that would accompany the appropriations bill reflecting the concerns of the subcommittees. On June 21, 1984, I had breakfast with Wally Berger during which he proposed a possible compromise between the House and Senate versions of the appropriations bills. I reported this to Jim Beggs, and we decided to ask Phil Culbertson to work with Berger and Malow to reach a satisfactory compromise version. The final version of the statement that appeared in the Conference Report contained both a provision for NASA to work on automation technology and also to study the unmanned option for a space station. The debate was over, the democratic process had worked its will, and we were now on our way.

XVII

Another Beginning

I have come to the end of my story. But, as happens so often, the end is really only another beginning. NASA could now start to implement the technical plans that John Hodge's task force had developed during the past two years. The program office to lead the space station effort in Washington was established on July 27, 1984, and Phil Culbertson was put in charge—just as we had planned in our conversations three years earlier. The program office at the Johnson Space Center was by now in full swing under the leadership of Mr. Neil Hutchinson. All of this was very positive. The technical plans were also coming along well. Culbertson had carried on a sometimes contentious and sometimes delicate negotiation with the directors of the various NASA centers to apportion the work on the space station so that people began to know just what they would be doing. And, some early possible configurations of the space station were beginning to emerge. There would be a central service module containing the basic systems necessary to operate the space station, power, guidance and control, and data handling. Then there would be the habitation modules in which people would live and work. The whole thing would be tied together with a structure, the construction of which would present technical problems of its own. Finally, planning was also in progress for the unmanned co-orbiting platforms that would accompany the space station.

What we had promised to deliver was this: we would build a permanently manned space station in low earth orbit, we would do it for something like $8 billion, we would deploy the principal elements of the space station in 1992, and we would execute the whole program in collaboration with our friends and allies abroad. This was the plan, and we now had the green light to go ahead with it. NASA had a new direction for the future.

There would be many problems; perhaps the most important ones were actually internal to NASA itself. Although the organizations to implement the space station program were put in place, the relationships between these and the other elements of the NASA organization were not well defined. The most important of these is the space shuttle program. We have succeeded in making the shuttle an engineering success, and we have demonstrated many of the operational capabilities inherent in the idea of a reusable spaceship. What we have not yet done is to demonstrate that the shuttle is an operational and an economic success. It is not yet clear that the customers of the shuttle have the necessary confidence in the system to use it the way it could be used. This is true both of the commercial customers and of the Air Force. It is also not clear whether the cost of operating the shuttle system is well understood as yet. There is a fine line between "improvement" for operational reasons and "improvement" to get a better system. At the present time, there is no clear way of distinguishing between these, and it may not be possible to make this distinction as long as the shuttle is operated by NASA.

The fact is that NASA is an engineering development organization —and a first-class one. The talent required for engineering development is substantially different than what is needed for operations. It may be that at some time in the next few years, a new operational organization will be established to run the shuttle. My own feeling is that the best way to do this is to follow the model we established in the 1960s when the communications satellites were "spun off" from NASA. What I am advocating is that a government-sponsored corporation be established to operate the space shuttle. This corporation, in the beginning at least, would be controlled by the government in that the government would own the majority of the stock. However, private investors would be encouraged to put money into the corporation. This is, of course, exactly what was done when the Communications Satellite Corporation (COMSAT) was established in 1962 to operate communications satellites. The interests of the military in the shuttle system would be accommodated by appropriate representation on the board of directors of the corporation and perhaps by requiring that some of the senior operating officers of the corporation be serving military officers. The two launch sites for the shuttle at the Kennedy Space Center and at the Vandenberg Air Force Base would be turned over to the new corporation. The mission control function for shuttle flights that is currently performed at the Johnson Space Center would be moved to the Kennedy Space Center. Eventually, it is possible that other launch vehicles in the inventory, or launch vehicles that are yet to be developed, will also be operated by the proposed corporation.

Until such an operational organization for the shuttle is established, NASA will continue to be the operational entity. The government will, of course, continue to pay for the shuttle operation—or for most of it—and will, therefore, be subsidizing the commercial users of the shuttle. There is ample precedent for such a policy. In the nineteenth century, the government also provided heavy subsidies for new transportation technologies that later on turned out to be extremely important both for civilian and military purposes. I am speaking here of railroads and steamships. The railroad technology was pushed very vigorously during the American Civil War, and after the war the government heavily subsidized the construction of railroads through massive land grants along the new railway rights of way. Steamship technology was subsidized by the U.S. Navy as a result of the experiences during the Mexican-American War in 1846–48. Both of these efforts at technology development were successful, and there is every reason to believe that a similar sequence of events will occur in the development of space transportation systems.

There will also be problems external to NASA. Perhaps the most important of these is the proper development of relationships with the military establishment. I have already discussed this problem elsewhere in this book, but it is worth a few more words. There is no question that the shuttle provides new capabilities that will be very important for future space operations related to the national security. Contrary to those people who fear the "militarization" of space or of the shuttle, these operations are much more likely to contribute to world stability rather than to increase the destructive potential of modern war. Space systems provide information for our national leaders that greatly reduce the uncertainties that they face in making critical decisions during times of crisis. Space systems are also absolutely essential for the monitoring and verification of arms control agreements. Therefore, even those people who are concerned about the military use of the shuttle should take these facts into account. The people in the military are, of course, quite legitimately concerned about the control of the launch vehicles that they must use to launch their most important payloads. This is, of course, the major reason why an organization must eventually be established to operate the space shuttle system that has the complete confidence of all the users of the shuttle, including the military.

There will also be continuing political problems. It is in the nature of our political system that decisions are not final until the programs to implement the decisions become a totally accepted part of the political scene. The debate over the space station that I have described in these pages will be repeated in a number of ways in the coming years. Although the current administration is thoroughly committed to the space station, this may not be true of the next one. Therefore, the persuasion process will once again have to be initiated in much the same way that it was when President Carter was persuaded to keep the space shuttle program on track in 1978 and 1979. Furthermore, NASA will have to go back to the Congress every year to secure funding for the space station program. During the debates that will be part of the future congressional authorization and appropriations processes, all of the issues that I have described will be aired again and reviewed. I expect that there will be members of the scientific community who will continue to oppose the construction of the space station. (There are still prominent members of the scientific community today who continue to voice their opposition to the space shuttle.) There will also continue to be opposition from other quarters—particularly the military establishment—for various reasons. Perhaps the most important publicly stated argument from all of these sources will continue to be that the money being spent on the space station could be spent more effectively on other aspects of our space program. In view of all of this, those of us who are convinced that the space station is indeed the proper next step in space must continue to be prepared to participate in these debates.

In the final analysis, the space station is a matter of having a vision for the future and of faith in that vision by the people. It is the function of the nation's leaders to create that vision. In 1803 President Jefferson made the decision to buy the Louisiana Territory from France. At the time he took this step, he had only the vaguest notion of what was in the vast territory. But Jefferson had vision, and he imagined that, at some time in the future, the United States would be a great continental nation rather than a small, fragmented group of former colonies strung along the eastern seaboard. In committing to the construction of the space station, President Reagan displayed the same kind of vision. We do not know exactly what we will ultimately do with a space station and where what we do will lead. This is not something that is really subject to quantitative analysis, and we should not believe that it is. What we do know is the lesson of history that whenever people have ventured into the unknown and have taken great risks, the horizons and the welfare of humanity have both been expanded. For my part, I am very grateful that I have had the opportunity to play a small part in this great enterprise.

XVIII

Tragedy and Tomorrow

On April 14, 1984, I made the decision to leave NASA and government service. Thus, I ended what was a little over fifteen years of working in the U.S. government after joining the NASA-Ames Research Center in February 1969. At the time, my reasons for taking this step seemed compelling. We had made the space shuttle fly and had developed the initial operational procedures. We had taken the next step in space by persuading President Reagan to propose the development and the deployment of a permanently occupied space station. These were significant achievements, and I felt satisfied to have had a part in bringing them about. However, I also felt that it was time to go and do something else. Prior to joining the federal government, I had spent many years teaching in a large state university, and I thought that I might be able to make some contributions by returning to that environment. Accordingly, I accepted an offer to become chancellor of the University of Texas System.

In addition to the professional reasons I have listed, there were personal ones as well. In my fifteen years of government service, I had accumulated personal debts of over $40,000, and these had to be paid. The job in Texas carried a salary that was well over twice what I was earning as deputy administrator of NASA, which made it possible to meet the obligations I had incurred. Also, my wife was not happy in Washington, and she wanted to leave whenever a good opportunity presented itself. So, the decision was made, and we drove from Washington to Austin in the middle of September 1984. It was a pleasant drive in perfect weather, and we were in excellent spirits when we arrived in Austin to take up our new duties. (I use the word "our" deliberately because in university posts of the type I was about to assume, the wife is very much part of the team.)

For the next eighteen months, I was heavily involved in learning my new job and becoming used to a new environment in Texas. All of my new friends in Texas and the new colleagues I was working with were extremely kind and helpful in putting us on the right track. I cut all my ties with NASA and with the Air Force, the two federal agencies I had worked for during my term of service in the federal government. I did this because I have seen too many old Washington hands remain on advisory committees to agencies that once employed them refighting old battles and telling war stories. I did not want to fall into that trap. However, I did retain two committee assignments that I felt were important and were not too closely related to what I had done in my almost eight years in Washington. One was a seat on the executive panel of the chief of naval operations and the other was membership on a committee chaired by Professor Frederick Seitz that was established to provide technical advice for the new director of the Strategic Defense Initiative Office, General Jim Abrahamson. These were both congenial assignments because they permitted me to retain good connections in the Department of Defense. I was especially pleased to continue the association with Jim Abrahamson since he had become a good friend during the years I spent in Washington.

On January 28, 1986, I was scheduled to address a group of senior active and retired Air Force officers who were meeting at Randolph Air Force Base in San Antonio. I drove to San Antonio in the morning with my old friend from Ames days, Mr. John W. Boyd, who had recently joined the University of Texas System as well. When we arrived at the gate, the air policeman told us that there had been an accident during the space shuttle launch scheduled for that morning. The public affairs people at the base had recorded the launch from the NASA television network (many of the Air Force bases are tied into this net), so we were able to take a close look at what had happened. Needless to say, we were horrified by what we saw. By playing the tape a number of times in slow motion, I could see that the vehicle yawed perceptibly just before the External Tank (ET) exploded. Initially, I thought that the small separation motors on the right-hand Solid Rocket Motor (SRM) had somehow fired prematurely, thus causing the vehicle to yaw and somehow pushing the SRM into the External Tank, which then led to

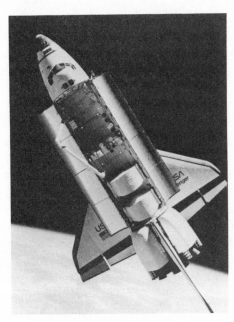

Challenger on
February 7, 1984.

the rupture of the tank and the subsequent explosion. As I sat in the
small conference room in the Base Public Affairs Office, the enormity
of what I was looking at on the television screen began to sink in. The
destruction of the *Challenger* in that terrible fireball would have far-
reaching consequences that were bound to be harmful to the American
space program and would probably nullify many of the things we had
worked so hard to achieve.

A few days later on January 31, 1986, there was a memorial service
for the crew that perished in the accident at the NASA-Lyndon B.
Johnson Space Center near Houston. It was one of the most difficult
days in my life to sit there and to share the grief. For me, it was not only
the tragedy of lives cut short and expectations that would not be fulfilled.
I also had the feeling that a world I had known had somehow come to
an end. It was a terrible experience to see my old friends in the Astronaut Corps sit grim-faced, listening to one of their own (Lt. Col.
Charles F. Bolden, USMC) deliver the eulogy. It was even harder to listen
to President Reagan sum up the whole tragedy in his usual eloquent
way and then promise that we would go on. I had no doubts about his
own commitment to our space program, but would he be able to

persuade his advisors, the Congress, and the American people to stick to the program even under adverse conditions?

Perhaps the most difficult moment of that most difficult day was the brief encounter I had with Jim Beggs, my old boss. Two months earlier, in December 1985, Beggs had been indicted for fraud against the government over a matter that dated back to the days when he was executive vice president of the General Dynamics Corporation. At the time of the accident, Beggs was on leave from NASA, a leave that had been granted by the president so that he could defend himself in the court case then pending. It must have been terribly difficult for him to endure these blows, first the indictment and then the accident, coming one after the other. When I saw him, all I could do was to embrace him and weep bitter tears.

On February 3, 1986, President Reagan appointed a commission headed by former Secretary of State William P. Rogers to investigate the accident. The commission was charged with finding the causes of the accident and then making appropriate recommendations that would prevent similar accidents in the future. The commission was given 180 days to prepare a report. By any standard, the commission performed its assignment in an exemplary manner both in developing the technical causes and in explaining what events led to the accident.

The space shuttle *Challenger* was destroyed seventy-three seconds after it lifted off from launch pad 39A at the Kennedy Space Center on January 28, 1986. The proximate cause of the accident was a failure in the seal of the joint between the lowest and next to the lowest segment in one of the Solid Rocket Motors of the vehicle. (This joint between the segments is called a "field joint" because it is made up in the "field," that is, at the Kennedy Space Center rather than at the manufacturer's plant.) That much has been established beyond any doubt. The real question is why the failure occurred. As with most accidents of this kind, there was a combination of factors that ultimately led to the accident, which started with the design of the seals and particularly the o-rings in the tang-and-clevis joint. There were some engineers knowledgeable in the area of seals and joints who, as early as 1977—four years before *Columbia*'s first flight—raised some questions regarding the design of the o-rings. (These objections are documented in the *Rogers Commission Report*, pp. 123–24.) At the time, a judgment was made that these objections were not serious enough to warrant changes, and so the design was retained. I was not in NASA in 1977 so I cannot speak from personal experience as to why NASA management chose not to heed these warnings. What is clear is that the seed for the ultimate tragedy nine years later had been sown.

My own part in the chain of events that led to the accident began when I returned to NASA early in 1981. I first became aware of the fact that we had a problem with the o-ring seals on the Solid Rocket Motor at the time when our engineering people were questioning whether these "field joints" on the SRM were really fail-safe. During the design of the space shuttle, an effort was made to make as many of the subsystems as possible "fail-safe." The idea was to design them in such a way that a single point failure would not have catastrophic consequences. In the case of the "field joints," this was accomplished by putting two o-rings in the joint on the theory that if the first one failed, then the second one would do the necessary job.

My memory is that questions as to whether the double o-ring system was really fail-safe began to be raised sometime in 1982. In February or March 1983 Mr. L. Michael Weeks, the deputy associate administrator of NASA for space flight, signed out a memorandum waiving the fail-safe requirements for the field joints in the Solid Rocket Motor. I remember discussing that matter with him at the time and concluding that such a step was justified. I argued at the time that we had more than a hundred successful firings of the Titan Solid Rocket Motor with a seal of somewhat similar design containing only one o-ring. I thought because of the Titan precedent that the risk of failure was small. As things turned out, this judgment was not correct because there are significant differences between the Titan and the SRM joints. I did not look at these differences with sufficient care at the time.

At the same time, we did have other serious problems with the shuttle vehicle that caused me to spend much more time on those issues than on the o-ring seals. There was, for example, the problem with the erosion of the Solid Rocket Motor nozzle. On one flight we came within a few seconds of burning out the nozzle before the rocket fuel itself was exhausted. I remember that we conducted a series of very detailed reviews to get to the bottom of this problem and then to fix it. We succeeded in attributing the problem to the failure of a third level contractor to follow the proper procedure in processing the heat resistant synthetic material that lines the nozzle. We also had difficulty with hydrogen leaks in the aft end of the vehicle. I remember that we postponed the first flight of *Challenger* for more than a month looking for hydrogen leaks in the aft engine compartment. We noticed this problem for the first time during the flight readiness firing of the *Challenger's* engines. We also had problems with the auxiliary power units on the Orbiter. These tended to run hot, and power failures were something that we always feared during every flight. All of these things were more important in my mind than the problem that I felt we had with the o-ring seals at the time.

The o-ring seal problem did gain my attention again just before I left NASA in 1984. On the tenth flight (STS-41B) we noticed some charring of the o-rings in the lower field joint. This phenomenon had been observed once before on the second flight (STS-2), but when it did not reappear we thought it was a one-time event. When we saw it again on the tenth flight, the question of what should be done was discussed at the Flight Readiness Review for the eleventh flight (STS-41C). After the completion of the Flight Readiness Review, I issued an "action item" asking for a complete review of all the Solid Rocket Motor seals and joints. My intention at the time was to review this problem in the same manner as we had done with the Solid Rocket Motor nozzles when we had problems with nozzle erosion. Unfortunately, this review was never held. I made the decision to leave NASA about two weeks after signing out the "action item" so the matter was apparently dropped. (The "action item" was issued on March 30, 1984, and I made the decision to leave NASA in mid-April 1984.) The due date for the review was May 30, 1984, and by that time I was a lame duck. I should have insisted on holding the review anyway. Perhaps that way, more attention would have been paid to the problem once I was gone.

The subsequent history is explained on page 132 of the *Rogers Commission Report*. The people at Marshall Space Flight Center and Thiokol decided that they would develop a plan to fix the o-ring problem rather than review the matter with the higher-level NASA management. It is apparently for this reason that nothing was done for fifteen months to make NASA management at the headquarters level more aware of the problems that were developing with the o-ring seals at the time. It is also for this reason, very probably, that Jim Beggs and other high-level officials at NASA could claim that they were not aware of any really serious problems with the o-ring seals (see page 135 of the *Rogers Com-*

mission Report). A complete review of the o-ring seal problems on the Solid Rocket Motors was finally held at NASA headquarters on August 19, 1985, fifteen months after the original request was made. Even then, the most senior person who attended the review was Mr. Weeks, a deputy associate administrator of NASA. Neither the administrator of NASA nor the associate administrator for space flight were present.

I was aware of the problem with the o-ring seals in 1983 and 1984. I felt that it would have to be dealt with as a result of the review I called for in a manner similar to the reviews we had for other problems we experienced, such as the Solid Rocket Motor nozzle erosion. At the time I did not think that the o-ring seal problem was as serious, for example, as the Solid Rocket Motor nozzle erosion problem or some of the problems we were having with the shuttle main engines. Nevertheless, I felt that the time had come to ask some serious questions about the technical situation.

A question that I have asked myself over and over again is whether I would have flown on that day. I was involved in the launch decision for twelve shuttle flights, so I have a little experience in how these things work. I cannot remember a single flight when some group of engineers who were responsible for one or another of the subsystems did not advise us to delay the launch. Sometimes we took their advice and postponed the launch, and at other times we went ahead and flew in spite of the advice we were given. The mere fact that a group of engineers opposed the launch because they were afraid one of their subsystems would not work was not enough to cancel the launch. In view of this, I do not know whether the recommendation of the Thiokol engineers not to fly would have been enough to persuade me not to launch *Challenger* on January 28, 1986.

I do have to confess that when I saw the pictures of the ice on the launch pad in the *Rogers Commission Report*, I was very surprised that the NASA management gave the go-ahead to fly. The launch pad structures and the gantry were both completely covered with ice, and there were a great many icicles. Those icicles become missiles when the pad vibrates during takeoff, and they can easily damage the vehicle. I was always especially concerned about damage to some of the tiles of the thermal protection system. The tubes that carry the liquid hydrogen for the regenerative cooling of the shuttle's main engine nozzles are also somewhat vulnerable to damage by flying ice. Obviously, a rupture of these tubes by flying ice could lead to catastrophe. We canceled launches with much less ice on the pad than I saw in those pictures during the years I had anything to say about whether or not to launch.

These are my impressions, but what I have said with respect to the decision to launch is clearly second guessing. I was not there, and so I will never know what I would have done. What is clear is that there was a failure in the human communication chain in the decision to launch as well as in the design and then in the flight readiness review process. The necessary information on which decisions should have been based apparently never reached top-level NASA management. In the case of the danger signals that we had of o-ring failures, the information did reach the top management level in March 1984 during the Flight Readiness Review for the eleventh launch (STS-41C). However, the review of the situation requested subsequent to the Flight Readiness Review that should have been conducted was not held. Thus, the problem was never properly explained to the top-level management (the administrator and his immediate assistants) because people were apparently under the impression that things were being fixed. This happened in spite of the fact that o-ring erosion problems much more serious than the ones experienced on the second and on the tenth flight were observed on flights after the twelfth flight (STS-41C). In the case of the decision to launch *Challenger* on the morning of January 28, 1986, the three senior people who bore the responsibility, the acting administrator of NASA, the associate administrator for space flight, and the director of the Space Shuttle Program did not know of the objections raised by the Thiokol engineers to the launch on January 28, 1986, because of the cold weather. This, to me, is evidence that the strong communication chain that should have existed between top management and the engineers in the field was not there.

It would be easy to conclude from this that all you have to do is to install a better information system and everything will be all right. Just make sure that all the working engineers report everything they are doing, and that will fix the problem. The situation is just not that simple. Setting up that kind of an information system would create more paper than could be followed. Much more would be written than read! The simple fact is that the people at the top of the organization must ask the right questions. It is a two-way street. The people in the ranks must tell the truth and promptly write clear reports. The people at the top must make judgments about which questions to ask, and they must have the proper instincts to ask the right ones. In my judgment, both the working engineers and the NASA leadership failed to meet these responsibilities on that fateful day and during the months that preceded it.

What comes next? Will we recover and continue along the path that was set for NASA when it was founded more than a quarter century ago? I believe that we will, but it will take time and effort to make the recovery. It is most important to fly the shuttle again as soon as possible. After a tragedy such as the one that befell *Challenger* and her crew, there is the danger that we will become too timid and establish so many hurdles that no one will ever again risk giving the order to launch. We must avoid this danger. As I write this in September 1986, I am concerned that the Russians will be able to gain very substantially both in actual flight experience and in shaping the public perception that they are the leading space faring nation in the world while we are waiting to fly again. We must not let that happen. I know that what I am saying here may go against the grain, but I believe that we may even want to accept some extra risks to fly the shuttle again sooner rather than later. We did this after the fire in the Apollo 204 capsule in 1967 in which three astronauts lost their lives. In that case, some of the planned test flights before the first lunar landing was scheduled were canceled so that the time lost due to the fire would be recovered. This involved some increased risks, but they were accepted. I believe that it may be necessary to do the same thing again if we are to restore public faith in NASA.

Fortunately, the new leaders of NASA understand these points as well as anyone in the world. Rear Admiral Richard H. Truly, one of the most experienced shuttle astronauts, was appointed NASA's associate administrator for space flight shortly after the accident. Truly is an experienced flier as well as a first-class engineer, and if anyone can get the shuttle flying quickly and safely, he is the man. Equally important is the appointment of Jim Fletcher to once again head NASA. There is no one better qualified to restore the nation's confidence in our ability to conduct space missions in a safe and efficient manner. In addition, he has all the necessary vision and experience to develop and articulate the plans that need to be made for NASA's future.

Since the accident, NASA's critics have had the upper hand in gaining the attention of the press and the public. This is inevitable given the circumstances, and it may even be important to air some of the issues again that have been described in the earlier chapters of this book. Was the overall program that was conceived in the late 1960s that called for the development of a reusable launch vehicle and the deployment of a permanently manned space station the right thing to do? If not, what changes should be instituted at the present time? Was the decision to convert all (or at least almost all) spacecraft to use the shuttle correct? The policy to put all (or almost all) space payloads on the shuttle had been controversial for a long time. It was actually modified to some extent even before the *Challenger* accident with the approval that the Air Force had secured to purchase ten large expendable launch vehicles based on the technology of the Titan rockets (these are called CELVs).

Will we now, because of the *Challenger* accident, pull everything off the shuttle and make only minimal use of the capability that it represents? I hope not, but there is no doubt that there are great pressures, technical, political, and economic, to do just that right now. It is too early to tell exactly how NASA's contribution to the American space program will evolve. I am optimistic that eventually we will regain our equilibrium and that NASA will resume the leadership role that it has held for so long.

For my own part, I now regret that I ever left NASA. I always realized that eventually we would lose one of the shuttles, since I knew from long experience that failure is an inevitable consequence of taking risks. I had hoped that the expected accident would happen later than it did, at a time when some of the operational organizations that I had advocated would be in place. However, all of this is speculation and probably not very useful. I do not know whether I could have done anything to prevent the loss of *Challenger* had I stayed on at NASA. What I know is that, somehow, I should have been there to share the grief and the pain with all my friends when the accident happened.

Appendixes

Appendix 1
Letter to Roy Jackson on Shuttle Technology, February 15, 1972

National Aeronautics and Space Administration
Ames Research Center
Moffett Field, California 94305

February 15, 1972

Mr. Roy P. Jackson
Code R
National Aeronautics and Space Administration
Washington, D.C. 20546

Dear Roy:

I am very sorry that I will not be able to attend the Shuttle Technology Assessment Meeting at Langley on Wednesday, February 16, for reasons that I explained to you in our telephone conversation. Because I believe this is a very important session I would like to put some of our thoughts on the shuttle down on paper for your use in your meeting with Dale Myers.

For the past months we have had a group working here at Ames assessing the state of readiness of the various different technologies that must come together and be made to work if a shuttle vehicle is to be successful. Last Saturday I met with this group and conducted a thorough review of their work. As you know, Ames has concentrated primarily on the problems related to the orbiter and my comments will therefore apply exclusively to this portion of the shuttle problem.

It is our considered opinion that there is no technical reason for not proceeding with the orbiter at the present time. This remark applies to any of the candidate orbiter configurations. The statement should also be qualified since some of the requirements that have been placed on the orbiter vehicle will have to be relaxed if the machine is to use only currently available technology. I will discuss some of these specific questions shortly. In the area of aerodynamics there seem to be no major technical problems. A number of the proposed configurations will work satisfactorily throughout the proposed flight regime. It seems to us that the orbiter aerodynamics problem is quite similar to other problems we are facing in the development of advanced aircraft such as the B-1. Even though there do not seem to be any technological unknowns in aerodynamics, considerable wind tunnel time will be required to complete the program. As you know, Ames has already contributed about 10,000 hours of wind tunnel time to the phase A and B parts of this program and we estimate that an even larger effort will be needed in the C and D phases of the project. Our best estimate of the wind tunnel time required for the program is about 10,000 hours per year for the next three years. In the first year of the phase B contractor support program Ames has contributed over half of all the work done in government facilities on contractor support and we presume that this trend will continue through phases C and D. There is one important difference between the phase A–B support effort and the work planned for phases C and D. In the first case we used primarily our smaller facilities such as the 6- by 6-foot wind tunnel but in the phases C and D support work our larger facilities will be needed to better simulate the Reynolds numbers encountered in the actual flight regime. This circumstance will approximately double the cost of the program per wind tunnel hour.

In the field of handling qualities, terminal guidance and the landing problem we expect that our flight simulators will play a major role. Although there are very significant differences between current jet transports and the space shuttle vehicle (primarily unpowered flight with the associated higher approach speeds) aircraft guidance and control technology in its current state can be used for landing the space shuttle vehicle on specially constructed runways. This includes the use of existing navigation aids such as VOR-DME, ILS and the microwave landing guidance system. Using this technology it is apparent to us that the orbiter vehicle can be guided to a successful landing after reentry from orbit. However, because of the unpowered aspect of flight it is not at all obvious that this can be done using conventional runways. It is our belief that flight investigations are required to assess the trade-offs in the use of more sophisticated avionics, spoilers, speed brakes, etc., and the runway width and length considering the effects of such external environmental phenomena as cross winds, gusts and fog. We will employ our simulators in assessing these trade-offs particularly the flight simulator for advanced aircraft to examine the approach and touch down dispersion problems in which pilot cues are critical. In addition the development of advanced airborne avionics systems being currently developed for the CV-90, and the microwave landing guidance system which is being installed at Crows Landing for our joint STOL program with the FAA give us avionics hardware with which to verify and extend the simulation investigations in flight. It is our belief that the combination of ground and airborne facilities that have been developed here will make it possible for us to advance the currently available technology to the point where the orbiter can indeed be landed on most conventional airports when shuttle operations commence. It is important to recognize that the avionics development, in our view, is probably not very sensitive to the particular configuration of the shuttle vehicle that is chosen. Consequently, the development of the avionics system and the development of the vehicle itself can probably proceed independently for sometime to come before the two must be integrated.

In the field of thermal protection it is our view that the shuttle orbiter can be flown using presently available ablator heat shield technology. There are some problems dealing with the bonding of the ablator to the metal panels and particularly with the inspection of the bonds to determine that they are indeed secure after manufacture. However, we do not believe this is an insuperable difficulty and we expect that ablating heat shields can be made to work on the shuttle orbiter vehicle. It should be recognized of course that the use of ablators means that the orbiters are not reusable in the general sense of the term. We believe therefore that in addition to the development of an ablator system we immediately undertake development of a thermal protection system using reusable surface insulators (RSIs). The state of technology for RSIs is not ready in our view at the present time. However, there is a good possibility that if a concentrated effort is made the RSI technology will be ready for use when the aerodynamic problems of the orbiter vehicle are solved. We expect to be very active in the development of thermal protection systems for the shuttle since our 60 megawatt arc facility will be the only one in which full scale thermal protection panels can be exposed to the actual conditions experienced in the reentry flight of the shuttle orbiter vehicle.

In reviewing the work we have done on the shuttle orbiter the accuracy of the statement you made at the last Management Council meeting became clearer to me than ever before. The development of the shuttle orbiter is largely a problem in aeronautics. The vehicle is a flying machine and the development program for the orbiter with respect to its aerodynamic configuration and properties will be very similar to the development programs for other advanced aircraft. The same is true in the area of guidance and control and avionics. It is clear to us therefore that a very large part of the orbiter development job will be done at OAST facilities through the contractors that win the competition for the construction of the vehicle. We feel strongly that Dale Myers must be made to clearly recognize this fact and to provide the necessary R&D support, preferably through your office. We estimate that at Ames we will have approximately 80 people working on the program for the next four years and that we will require R&D funding of the order of four million dollars per year. We feel strongly that a vigorous statement of these requirements for OAST is the single most important point that should be made during the meeting with Dale Myers on Shuttle Technology Assessment on Saturday, February 19. The shuttle orbiter is essentially a classical aeronautics problem; it should be recognized as such and dealt with accordingly.

With best personal regards,

Sincerely yours,

Hans Mark

cc:
OAST Management Council
A. O. Tischler, NASA Hq.

Appendix 2
Unclassified Version of PD-42, October 10, 1978

The White House
Washington

October 10, 1978

Presidential Directive/NSC-42

TO: The Secretary of State
The Secretary of Defense
The Secretary of the Interior
The Secretary of Agriculture
The Secretary of Commerce
The Secretary of Energy
The Director, Office of Management and Budget
The Assistant to the President for Domestic Affairs and Policy
The Administrator, Agency for International Development
The Director, Arms Control and Disarmament Agency
The Chairman, Joint Chiefs of Staff
The Director of Central Intelligence
The Administrator, National Aeronautics and Space Administration
The Director, Office of Science and Technology Policy
The Director, National Science Foundation

SUBJECT: Civil and Further National Space Policy

[First paragraph intentionally deleted.]

ADMINISTRATION CIVIL SPACE POLICY. The United States' overarching civil space policy will be composed of three basic components.

First: Space activities will be pursued because they can be uniquely or more efficiently accomplished in space. Our space policy will become more evolutionary rather than centering around a single, massive engineering feat. Pluralistic objectives and needs of our society will set the course for future space efforts.

Second: Our space policy will reflect a balanced strategy of applications, science, and technology development containing essential key elements that will:

—Emphasize applications that will bring important benefits to our understanding of earth resources, climate, weather, pollution, and agriculture.

—Emphasize space science and exploration in a manner that permits the nation to retain the vitality of its space technology base, yet provides short-term flexibility to impose fiscal constraints when conditions warrant.

—Take advantage of the flexibility of the Space Shuttle to reduce operating costs over the next two decades.

—Increase benefits by increasing efficiency through better integration and technology transfer among the national programs and through more joint projects.

—Assure US scientific and technological leadership for the security and welfare of the nation and to continue R&D necessary to provide the basis for later programmatic decisions.

—Provide for the private sector to take an increasing responsibility in remote sensing and other applications.

—Demonstrate advanced technological capabilities in open and imaginative ways having benefit for developing as well as developed countries.

—Foster space cooperation with nations by conducting joint programs.

—Confirm our support for the continued development of a legal regime for space that will assure its safe and peaceful use for the benefit of all mankind.

Third: It is neither feasible nor necessary at this time to commit the US to a high-challenge, highly-visible space engineering initiative comparable to Apollo. As the resources and manpower requirements for Shuttle development phase down, we will have the flexibility to give greater attention to new space applications and exploration, continue programs at present levels, or contract them. An adequate Federal budget commitment will be made to meet the objectives outlined above.

SPACE APPLICATIONS. The President has approved the following:

Government Role in Remote Sensing
1. *Land Programs.* Experimentation and demonstration will continue with LANDSAT as a developmental program. Operational uses of data from the experimental system will continue to be made by public and private users prepared to do so. Strategies for the future of our civil remote sensing efforts are to be addressed in the FY 1980 budget review. This review should examine approaches to permit flexibility to best meet the appropriate technology mix, organizational arrangements, and potential to involve the private sector.

2. *Integrated Remote Sensing System.* NASA will chair an interagency task force to examine options for integrating current and future potential systems into an integrated national system. This review will cover technical, programmatic, private sector, and institutional arrangements. Emphasis will be placed on user requirements; as such, agency participation will include Commerce, Agriculture, Interior, Energy, State, appropriate Executive Office participation, as well as Defense, the DCI, and others as appropriate. This task force will submit recommendations to the Policy Review Committee (Space) by August 1, 1979, for forwarding to the President prior to the FY 1981 budget review.

3. *Weather Programs.* In the FY 1980 budget review, OMB—in cooperation with Defense, the DCI, NASA, and NOAA—will conduct a cross-cut review of meteorological

satellite programs to determine the potential for future budgetary savings and program efficiency. Based on this cross-cut, the Policy Review Committee (Space) will assess the feasibility and policy implications of program consolidation by April 1, 1979.

4. *Ocean Programs.* Any proposed FY 1980 new start for initial development of a National Satellite System (NOSS) will be reviewed based on a ZBB priority ranking. The Policy Review Committee (Space) will assess the policy implications of combining civil and military programs as part of this process.

5. *Private Sector Involvement.* Under the joint chairmanship of Commerce and NASA, along with other appropriate agencies, a plan of action will be prepared by February 1, 1979, on how to encourage private investment and direct participation in the establishment and operations of civil remote sensing systems. NASA and Commerce jointly will be the contacts for the private sector on this matter and will analyze proposals received before submitting to the Policy Review Committee (Space) for consideration and action.

[Paragraph intentionally deleted.]

Communications Satellite R&D. NASA will undertake carefully selected communications technology R&D. The emphasis will be to provide better frequency and orbit utilization approaches. Specific projects selected will compete with other activities in the budget process.

Communications Satellite Services. Commerce's National Telecommunications and Information Administration (NTIA) will formulate policy to assist in market aggregation, technology transfer, and possible development of domestic and international public satellite services. This policy direction is intended to stimulate the aggregation of the public service market and for advanced research and development of technology for low-cost services. Under NTIA this effort will include: (a) an identified 4-year core budget for Commerce to establish a management structure—competitive against other budgetary priorities in Commerce—to purchase bulk services for domestic and international use; (b) support for advanced R&D on technologies to serve users with low-volume traffic requirements subject to its competitiveness coordination with NTIA in translating domestic experience in emerging public service programs into potential programs for lesser-developed countries and remote territories.

Long-term Economic Activity. It is too early to make a commitment to the development of a satellite solar power station or space manufacturing facility. There are very useful intermediate steps that would allow the development and testing of key technologies and experience in space industrial operations without committing to full-scale projects. We will pursue an evolutionary program to stress science and basic technology —integrated with a complementary ground program—and will continue to evaluate the relative costs and benefits of proposed space activities compared to earth-based activities.

SPACE SCIENCE AND EXPLORATION GOALS
Priorities at any given time will depend upon the promise of the science, the availability of particular technology, and the budget situation in support of the following Presidentially-approved goals:

—We will maintain US leadership in space science and planetary exploration and progress.

—The US will continue a vigorous program of planetary exploration to understand the origin and evolution of the solar system. Our goal is to continue the reconnaissance of the outer planets and to conduct more detailed exploration of Saturn, its moons, and its rings; to continue comparative studies of the neighboring planets, Venus and Mars; and to conduct reconnaissance of comets and asteroids.

—To utilize the space telescope and free-flying satellites to usher in a new era of astronomy, as we explore interstellar molecules, quasars, pulsars, and black holes to expand our understanding of the universe and to complete the first all sky survey across the electromagnetic spectrum.

—To develop a better understanding of the sun and its interaction with the terrestrial environment. Space probes will journey towards the sun. Earth orbiting satellites will measure the variation in solar output and determine the resultant response of the earth's atmosphere.

—To use the Space Shuttle and Spacelab, in cooperation with the Western Europeans, to conduct basic research that complements earth-based life science investigations and human physiology research.

—Our policy in international space cooperation should include three primary elements: (1) support the best science available regardless of national origin, but expand our international planning and coordinating effort; (2) seek supplemental foreign support only for selected experiments-spacecraft which have been chosen on the basis of sound scientific criteria; and (3) avoid lowering cooperative activities below the threshold where our science and internation cooperative efforts would suffer.

STEPS TO INCREASE BENEFITS FOR RESOURCES EXPENDED. The President has approved the following:

Strategy to Utilize the Shuttle
1. [Paragraph intentionally deleted.]
2. [Paragraph intentionally deleted.]
3. Incremental improvements in the Shuttle transportation system will be made as they become necessary and will be examined in the context of emerging space policy goals. An interagency task force will make recommendations on what future capabilities are needed. Representation will include NASA, Defense, the DCI, Commerce, Interior, Agriculture, OMB, NSC, OSTP, State and others as appropriate. This task force will submit the findings to the Policy Review Committee (Space) for transmittal to the Presidency by August 1, 1979.
4. [Paragraph intentionally deleted.]

Technology Sharing. The existing Program Review Board (PRB) will take steps to enhance technology transfer between the sectors. The objective will be as directed in PD/NSC-37, to maximize efficient utilization of the sectors while maintaining necessary security and current management relationships among the sectors. The PRB will submit an implementation plan to the Policy Review Committee (Space) by May 15, 1979. In addition, the PRB will submit subsequent annual progress reports.

(Signed)

Zbigniew Brzezinski

Appendix 3
"USAF's Three Top Priorities," by the Honorable
Hans M. Mark, Secretary of the Air Force,
Article from *Air Force Magazine*, September 1979

In the coming decade, the most important test our country will face is to maintain the position we have as a major world power. The effort necessary to achieve this objective will have to do with the expansion and modernization of our industrial capacity, major improvements in productivity by the application of new technology, and, most important of all, our ability to achieve some independence from foreign supplies of oil. Much of this effort will be carried on by our traditional private-enterprise system, and the personal initiative of individual people will play an important part.

The military component of this effort will be crucial. We must continue the ability to deter our principal military competitor, the Soviet Union, from launching a major strategic attack on this country. We must continue to retain the ability to respond to military contingencies around the world in which our vital interests might be involved. Finally, as a world power, we need to continue developing a worldwide perspective on the interests that we have elsewhere. It is entirely fitting that even in developing our military posture, private, voluntary, civilian organizations such as the Air Force Association should play a vital part. This is in the long tradition of our military history to minimize the distinction between military and civilian pursuits.

There are three important priorities for the Air Force as the effort to maintain our world power status develops:

1. The enhancement of our strategic forces to maintain a level that will ensure strategic equivalence with the Soviet Union.

2. The enhancement of strategic and tactical airlift so that we can adequately respond to worldwide contingencies where our national interests are involved.

3. The development of a doctrine and an organization that will permit greatly increased Air Force activities in space in order to take advantage of new technology to enhance communications, reconnaissance, and other vital Air Force functions.

The modernization of the strategic deterrent is our first priority today. For the past fifteen years, the Soviet Union has consistently increased its strategic power and is now in a position where in a number of fields it could have superiority. The current balance could easily be upset if we do not take aggressive steps to reverse present trends. We have upgraded our Minuteman ICBM force, and some of our B-52s are being modified to carry the new air-launched cruise missile. The President has approved our plans to deploy a new missile system, and we are proceeding to improve our strategic surveillance and strategic command and control systems.

These actions fall short of a complete solution. They must be followed by a basing decision for the MX, positive plans for a new bomber, and continual real growth in the defense budget. All of these steps must be taken to correct projected vulnerabilities and to provide a secure strategic force structure to support further negotiations for strategic arms limitations. These actions will help stabilize the strategic relationship between ourselves and the Soviet Union, but they will only work if we have a position of equivalence with our major military opponent.

We must continue to be able to respond to crisis situations in different parts of the world where our vital interests are involved. We must be able to supply allies rapidly, and we must be able to project forces if that needs to be done. We must also be able to evacuate rapidly Americans who are endangered in revolutions or conflicts around the world. While we have good strategic airlift capability, we do not have enough aircraft, and the tactical airlift force will have to be modernized in the next decade. Thus, airlift modernization must be considered as an essential element of our overall ability to operate as a major power, and a renewed effort must be made to improve this portion of our force.

The United States today is operationally dependent on space systems for a variety of functions that are of critical importance to our national security. Our military forces today have improved, more reliable communications, more accurate maps, more exact navigation aids, better weather information, and the capability for more reliable and quicker warning of attack due to a network of satellites that provide these things. In the next few years, the Space Shuttle will start flying, and once this happens our ability to operate in space will be greatly increased. We must see to it now that we have the organization and the doctrinal concepts that will permit us to take advantage of this new technical capability.

There are other things that also have to be done. I am speaking here particularly of the need to make it more attractive for young people to join and remain in our military forces. We also need to make certain that we can fly the airplanes we have, which means that more attention needs to be paid to maintenance and spare parts replenishment. However, I cannot dwell here on everything that is important.

It is important to recognize that the Air Force Association is a major factor in maintaining our connection with the civilian world and in providing a voice for the Air Force in many arenas where it would not otherwise be heard. I look forward very much to working with you to enhance the ability of the Air Force to help defend our country.

Appendix 4
Mark and Silveira, "Notes on Long Range Planning," August 1981

The development of long range planning for NASA should be based on Section 5 in the 1958 Space Act requiring "the preservation of the role of the United States as a leader in aeronautical and space science and technology. . . ." This may be a difficult thing to do in view of limited funds that will be available to NASA in the coming years but the intent of the statement in the law is crystal clear and NASA must act accordingly.

(1) FACILITIES
Fundamental to all that NASA does are the facilities that exist at NASA's research and development Centers. It is not always recognized but the NASA aeronautical facilities are vital, not only to aircraft design, but also to the development of our space technology. For example, the 40' x 80' wind tunnel at the Ames Research Center which is justified solely as an aeronautical facility was used for testing the flying qualities of the Space Shuttle during the critical approach and landing phase. A one-third scale model was tested many hundreds of hours in the wind tunnel to insure performance, stability, and control characteristics. There are many other examples where wind tunnel and high temperature facilities are used to insure safe flight of a spacecraft as it passes through the atmospheric portion of its flight.

Broadly speaking, NASA's facilities fall into five separate categories:
1. Wind Tunnels
2. Flight and Operations Simulators
3. Propulsion Test Facilities
4. Experimental Airplanes
5. Computational Facilities

Recently, heavy investments have been made in required wind tunnel facilities. Approximately $250M have been spent, improving the Ames 40' × 80' wind tunnel and building at Langley the High Reynolds Number Cryogenic Tunnel. Large investments have also been made in flight simulators, although more needs to be done in developing and building simulators to overcome current deficiencies. There is a need to develop more facilities for the simulation of operations and construction in space with a zero "g" environment and under demanding thermal conditions. The major aeronautical propulsion facility in the country is being developed by the United States Air Force at the Arnold Engineering Development Center. NASA must take advantage of this facility as best it can. NASA must also develop a policy toward the development of propulsion facilities at the Lewis Research Center. Particularly, NASA must also see to it that the rocket propulsion test stands are adequate for programs in launch vehicles that may be initiated following the completion of the Space Shuttle program. Experimental aircraft tend to be more specialized toward specific flight configurations. However, there are some programs such as the F-8 fly-by-wire aircraft and the Boeing 737 control configured vehicle in which the aircraft are used more-or-less as general purpose simulation facilities. Computers are not usually regarded as facilities but they should be viewed as such. The Numerical Aerodynamic Simulator now being proposed is particularly important in this regard since it may overcome certain limitations in the simulations of the other facilities now operated by NASA (wind tunnels, propulsion facilities and flight simulators) if the promise of computational methods in aerodynamics, chemically reacting flows, and dynamic structures can be realized. The maintenance and development of the necessary facilities to accomplish the mission stated in the law must, therefore, have the highest institutional priority in NASA.

(2) AERONAUTICS

Work in aeronautics by NASA, and the NACA prior to 1958, has traditionally been oriented toward the support of military and civil aviation. Future interest in the military is likely to be centered on the development of a new long range combat aircraft (LRCA) by the United States Air Force having low radar, infrared, and visible observables (i.e., stealth technology), the creation of a new family of V/STOL aircraft for the Navy, and the continuing enhancement of the performance of rotor craft for the Army. To maintain a lead in civil aircraft sales, continual improvements must be made for greater economy. The technology suitable for third level carriers (i.e., commuter airlines) is likely to be the major civil requirement. The latter is especially important in view of the inroads being made by foreign competition in that field. Right now the Dehavilland Twin Otter, the DHC-7, and the Shorts Skyvan dominate that field in the United States. In addition to all of these things, a strong basic research program in fluid mechanics, materials and other topics related to aeronautics and space vehicles must be maintained.

(3) THE SPACE SHUTTLE

The major technological development carried out by NASA in the last decade is the Space Shuttle vehicle. That basic development is now nearly complete and the next step is to turn it into an operational system. This effort must have the highest programmatic priority in NASA for the coming years to realize a return for this large investment. It should take about three years to make the Space Shuttle an operational transportation system. It is necessary to arrive at an agreed-upon definition of what is meant by "operational" and to determine whether NASA should be the agency that operates the Shuttle or whether some other institutional mechanism needs to be provided for that purpose. The organizational structure needs to be developed for Shuttle Operations. No matter how the matter of Shuttle Operations is finally decided, the Johnson Space Center should phase out of the operational mission during the next three years. It is very unlikely that it will be possible to control costs of operations if the developmental attitudes that prevail at the Johnson Space Center dominate after the Space Shuttle becomes operational. The operations of the Space Shuttle, both launch as well as mission control, should be handled by the Kennedy Space Center and by Vandenberg Air Force Base once the West Coast launch facility is complete.

(4) THE SPACE STATION

While the Space Shuttle becomes operational, a project to establish a permanent presence in space (i.e., a Space Station) should be initiated. This should become the major new goal of NASA and, some time during the next two years, the President should be persuaded to issue a statement proclaiming a national commitment to that effect. The necessary arguments that justify this step must be carefully developed now, and these arguments range from national security (i.e., arms control verification, military surveillance) to the improvement of space operations (i.e., satellite maintenance on orbit and other things of this kind). The necessary committees of the National Academy of Engineering, the National Academy of Sciences, and other bodies of this kind should be established to set up now the technical baselines for this new enterprise.

(5) UNMANNED LAUNCH VEHICLES

The Shuttle program has led to the creation of a new propulsion technology which should be further exploited. It is now generally agreed that unmanned launch vehicles will not be phased out completely once the Shuttle is operational. They will always be necessary to supplement the Shuttle launch capability. The current launch vehicles (Atlas, Titan, Delta) are based on technology that is now thirty years old and should be replaced by more efficient and economical vehicles. New unmanned launch vehicles based on the Shuttle technology using solid rocket boosters and the Shuttle's main engine system should be developed. The solid rocket booster itself is an excellent rocket with sea level thrust of the order of 2.5M lbs. Several solid rocket boosters strapped together could provide a formidable launch vehicle in terms of payload capacities. Such a vehicle with three solid rocket boosters could put into low earth orbit a payload weighing something like 100,000 lbs. and perhaps up to 20,000 lbs. into geosynchronous orbit with an appropriate upper stage. An important feature of the solid rocket booster is that they are recoverable which means that the cost advantages inherent in that property could be important. This new generation of launch vehicles would not be "expendable" although it would be unmanned.

(6) SCIENTIC EXPLORATION

NASA has a fundamental responsibility to continue with the scientific exploration of objects in space and conditions in space. In the coming decade, scientific investigations conducted in earth orbit will be the most important because these take the best advantage of the unique properties of the Shuttle. Specifically, this means that astronomy, experiments involving certain cosmological things such as general relativity and experiments in zero gravity using Spacelab will be the dominant trend in scientific space research. An extremely important aspect of this are the medical and biological experiments to be done using the Shuttle to establish what must be done to permit people, animals, and plants to live in zero gravity conditions for lengthy periods. It is probable that planetary exploration will be deemphasized somewhat until we have a Space Station that can serve as a base for the launching of a new generation of planetary exploration spacecraft. It is apparent that the return of samples from various bodies in the solar system will be given highest priority once that time arrives.

(7) SPACE APPLICATIONS

The applications program should emphasize the scientific part of earth observations, specifically oceanography, geodesy, and things of this kind. In view of the Administration's policies with respect to technical demonstration programs, NASA should de-emphasize efforts to commercialize various applications projects. The applications program should also emphasize technology development and should cooperate closely with the national security community in these efforts. It is likely that the nation's surveillance satellites will move to geosynchronous orbit in the next two decades. This means that large space structures will be required, mirrors, antennas, and other systems of this type. NASA should be extremely active in the development of this technology and should establish the closest possible support of the national security community in achieving these objectives.

A few thoughts regarding future directions for NASA have been outlined in this paper. Obviously, much more detail needs to be done to develop some of these ideas. It is very important to begin now by setting up the proper procedural methods within NASA as well as the NASA advisory structure to make certain that these ideas are properly considered and developed into a coherent long range plan for the nation's aeronautical and space programs.

Hans Mark
Milton Silveira
August 1981

TO: AD/Dr. Mark
OK, let's go.
James M. Beggs.
Sept. 24, 1981

[Note: This was the response to the planning document by the administrator.]

Appendix 5
Letter to Dr. James C. Fletcher Setting Up His Committee, September 18, 1981

National Aeronautics and Space Administration
Washington, D.C. 20546

September 18, 1981

Dr. James C. Fletcher
Burroughs Corporation
7726 Old Springhouse Road
McLean, VA 22102

Dear Jim:

This letter will confirm the intent of the conversation which we had earlier this week in which I asked that you assist us in the preparation of a plan which will lead to the development of—for temporary lack of a better name—the United States Space Station.

Both Mr. Beggs and I believe that NASA has reached the point where it is necessary to become more specific in our planning of major agency initiatives to be undertaken during the next three or four years. We also believe that the most significant of these should be the development of a system which will permit men—and women—to live and work in space for periods of time measured at least in months. Such a facility was, as you recall, considered to be the logical follow-on and companion to the Shuttle at the time that the Shuttle was conceived, now ten or twelve years ago.

I would like for you to assemble a group of six to eight individuals to counsel with us on the technical, economic, political, and bureaucratic factors to be considered as we develop the plan which will lead to the initiation and implementation of the project. As you recall, the individuals that came to mind during our conversation included Willis Shapley, Tom Paine, Frank Lehan, Jim Arnold, and Les Dirks. As a first step, I think it would be appropriate to bring this group together for a relatively unstructured brainstorming session which would expose issues, questions and problems which would receive the benefit of off-line considerations and preparations for subsequent meetings. I am not prepared at this moment to suggest either form or scope for the task which I am asking you to undertake, but I think that will evolve rather naturally as you give it some consideration and we have some further conversations.

Jim and I have asked Phil Culbertson to undertake the responsibility for directing the NASA activities toward the definition and approval of this project. I therefore believe that it would be both desirable and appropriate for him to be an ex officio member of your group.

As you know, Jay Keyworth, in his responsibility as Director of the Office of Science and Technology Policy, is directing a reexamination of existing space policy with an initial emphasis on Shuttle and other launch vehicle systems and capabilities. He plans to complete the first phase of the study; i.e., those factors which could affect the 1983 budget, by December of this year. This next major program of NASA is clearly related to these studies and I therefore think that it is desirable for us to aim toward a preliminary definition of plans and strategies in the same time frame. I would like to defer for the time being a decision about other dates which we can establish as we more precisely define the scope of the undertaking.

I would like for you and Phil to meet with me in a couple of weeks after each of us has had a little more time to flesh out our thoughts. In the meantime, you should feel free to contact those people about whom we talked. Phil will contact you to make arrangements for our next discussion.

I appreciate your willingness to undertake this task, Jim, and wouldn't, as I am sure you are aware, have imposed on you and your time if I didn't feel that it is of vital importance, not only to NASA, but more importantly, to the United States.

Sincerely
Hans Mark
Deputy Administrator

NASA TASK FORCE

Chairman, Dr. James C. Fletcher, Burroughs Corporation

Dr. George E. Mueller, Chairman of the Board and President, System Development Corporation

Mr. Lesley C. Dirks, Deputy Director for Science and Technology, Central Intelligence Agency

Dr. Robert R. Gilruth

Mr. Willis H. Shapley

Dr. James Arnold, Department of Chemistry, University of California, San Diego

Mr. Frank W. Lehan

Dr. Thomas O. Paine, President and Chief Operating Officer, Northrop Corporation

Mr. William Anders, Vice President and General Manager, General Electric Company

Lt. Gen. Pete Crow (USAF, Ret.)

Appendix 6
Reagan Administration Space Policy Statement,
July 4, 1982

The White House
Washington

July 4, 1982

National Space Policy

The President announced today a national space policy that will set the direction of US efforts in space for the next decade. The policy is the result of an interagency review requested by the President in August 1981. The ten-month review included a comprehensive analysis of all segments of the national space program. The primary objective of the review was to provide a workable policy framework for an aggressive, farsighted space program that is consistent with the Administration's national goals.

As a result, the President's Directive reaffirms the national commitment to the exploration and use of space in support of our national well-being, and establishes the basic goals of United States space policy which are to:

— strengthen the security of the United States;

— maintain United States space leadership;

— obtain economic and scientific benefits through the exploitation of space;

— expand United States private sector investment and involvement in civil space and space related activities;

— promote international cooperative activities in the national interest; and

— cooperate with other nations in maintaining the freedom of space for activities which enhance the security and welfare of mankind.

The principles underlying the conduct of the United States space program, as outlined in the Directive are:

— The United States is committed to the exploration and use of space by all nations for peaceful purposes and for the benefit of mankind. "Peaceful purposes" allow activities in pursuit of national security goals.

— The United States rejects any claims to sovereignty by any nation over space or over celestial bodies, or any portion thereof, and rejects any limitations on the fundamental right to acquire data from space.

— The United States considers the space systems of any nation to be national property with the right of passage through and operation in space without interference. Purposeful interference with space systems shall be viewed as an infringement upon sovereign rights.

—The United States encourages domestic commercial exploitation of space capabilities, technology, and systems for national economic benefit. These activities must be consistent with national security concerns, treaties and international agreements.

—The United States will conduct international cooperative space-related activities that achieve scientific, political, economic, or national security benefits for the nation.

—The United States space program will be comprised of two separate, distinct and strongly interacting programs—national security and civil. Close coordination, cooperation and information exchange will be maintained among these programs to avoid unnecessary duplication.

—The United States Space Transportation System (STS) is the primary space launch system for both national security and civil government missions. STS capabilities and capacities shall be developed to meet appropriate national needs and shall be available to authorized users—domestic and foreign, commercial and governmental.

—The United States will pursue activities in space in support of its right of self-defense.

—The United States will continue to study space arms control options. The United States will consider verifiable and equitable arms control measures that would ban or otherwise limit testing and deployment of specific weapons systems, should those measures be compatible with United States national security.

SPACE TRANSPORTATION SYSTEM

The Directive states that the Space Shuttle is to be a major factor in the future evolution of United States space programs, and that it will foster further cooperative roles between the national security and civil programs to insure efficient and effective use of national resources. The Space Transportation System (STS) is composed of the Space Shuttle, associated upper stages, and related facilities. The Directive establishes the following policies governing the development and operation of the Space Transportation System:

—The STS is a vital element of the United States space program, and is the primary space launch system for both United States national security and civil government missions. The STS will be afforded the degree of survivability and security protection required for a critical national space resource. The first priority of the STS program is to make the system fully operational and cost-effective in providing routine access to space.

—The United States is fully committed to maintaining world leadership in space transportation with a STS capacity sufficient to meet appropriate national needs. The STS program requires sustained commitments by each affected department or agency. The United States will continue to develop the STS through the National Aeronautics and Space Administration (NASA) in cooperation with the Department of Defense (DOD). Enhancement of STS operational capability, upper stages and methods of deploying and retrieving payloads should be pursued, as national requirements are defined.

—United States Government spacecraft should be designed to take advantage of the unique capabilities of the STS. The completion of transition to the Shuttle should occur as expeditiously as practical.

—NASA will assure the Shuttle's utility to the civil users. In coordination with NASA, the DOD will assure the Shuttle's utility to national defense and integrate national security missions into the Shuttle system. Launch priority will be provided for national security missions.

—Expendable launch vehicle operations shall be continued by the United States Government until the capabilities of the STS are sufficient to meet its needs and obligations. Unique national security considerations may dictate developing special purpose launch capabilities.

—For the near term, the STS will continue to be managed and operated in an institutional arrangement consistent with the current NASA/DOD Memoranda of Understanding. Responsibility will remain in NASA for operational control of the STS for civil missions and in the DOD for operational control of the STS for national security missions. Mission management is the responsibility of the mission agency. As the STS operations mature, the flexibility to transition to a different institutional structure will be maintained.

—Major changes to STS program capabilities will require Presidential approval.

THE CIVIL SPACE PROGRAM

In accordance with the provisions of the National Aeronautics and Space Act, the Directive states that the civil space program shall be conducted:

—to expand knowledge of the Earth, its environment, the solar system and the universe;

—to develop and promote selected civil applications of space technology;

—to preserve the United States leadership in critical aspects of space science, applications and technology; and

—to further United States domestic and foreign policy objectives.

The Directive states the following policies which shall govern the conduct of the civil space program:

—United States Government programs shall continue a balanced strategy of research, development, operations, and exploration for science, applications and technology. The key objectives of these programs are to: (1) preserve the United States preeminence in critical space activities to enable continued exploitation and exploration of space; (2) conduct research and experimentation to expand understanding of: (a) astrophysical phenomena and the origin and evolution of the universe through long-lived astrophysical observation; (b) the Earth, its environment, its dynamic relation with the Sun; (c) planetary, and lunar sciences and exploration; and (d) the space environment and technology to advance knowledge in the biological sciences; (3) continue to explore the requirements, operational concepts, and technology associated with permanent space facilities; (4) conduct appropriate research and experimentation in advanced technology and systems to provide a basis for future civil applications.

—The United States Government will provide a climate conducive to expanded private sector investment and involvement in space activities, with due regard to public safety and national security. These space activities will be authorized and supervised or regulated by the government to the extent required by treaty and national security.

—The United States will continue cooperation with other nations in international space activities by conducting joint scientific and research programs, consistent with technology transfer policy, that yield sufficient benefits to the United States, and will support the public, nondiscriminatory direct readout of data from Federal civil systems to foreign ground stations and the provision of data to foreign users under specified conditions.

—The Department of Commerce, as manager of Federal operational space remote sensing systems, will: (1) aggregate Federal needs for these systems to be met by either the private sector or the Federal government; (2) identify needed research and development objectives for these systems; and (3) in coordination with other departments or agencies, provide regulation of private sector operation of these systems.

THE NATIONAL SECURITY SPACE PROGRAM

The Directive states that the United States will conduct those activities in space that it deems necessary to its national security. National security space programs shall support such functions as command and control, communications, navigation, environmental monitoring, warning, surveillance and space defense. The Directive states the following policies which shall govern the conduct of the national security program:

—Survivability and endurance of space systems, including all system elements, will be pursued commensurate with the planned use in crisis and conflict, with the threat, and with the availability of other assets to perform the mission. Deficiencies will be identified and eliminated, and an aggressive, long-term program will be undertaken to provide more-assured survivability and endurance.

—The United States will proceed with development of an anti-satellite (ASAT) capability, with operational deployment as a goal. The primary purposes of a United States ASAT capability are to deter threats to space systems of the United States and its Allies and, within such limits imposed by international law, to deny any adversary the use of space-based systems that provide support to hostile military forces.

—Security, including dissemination of data, shall be conducted in accordance with Executive Orders and applicable directives for protection of national security information and commensurate with both the missions performed and the security measures necessary to protect related space activities.

INTER-PROGRAM RESPONSIBILITIES

The Directive contains the following guidance applicable to and binding upon the United States national security and civil space programs:

—The national security and civil space programs will be closely coordinated and will emphasize technology sharing within necessary security constraints. Technology transfer issues will be resolved within the framework of directives, executive orders, and laws.

—Civil Earth-imaging from space will be permitted under controls when the requirements are justified and assessed in relation to civil benefits, national security, and foreign policy. These controls will be periodically reviewed to determine if the constraints should be revised.

—The United States Government will maintain and coordinate separate national security and civil operational space systems when differing needs of the programs dictate.

POLICY IMPLEMENTATION

The Directive states that normal interagency coordinating mechanisms will be employed to the maximum extent possible to implement the policies enunciated. A Senior Interagency Group (SIG) on Space is established by the Directive to provide a forum to all Federal agencies for their policy views, to review and advise on proposed changes to national space policy, and to provide for orderly and rapid referral of space policy issues to the President for decisions as necessary. The SIG (Space) will be chaired by the Assistant to the President for National Security Affairs and will include the Deputy Secretary of Defense, Deputy Secretary of State, Deputy Secretary of Commerce, Director of Central Intelligence, Chairman of the Joint Chiefs of Staff, Director of the Arms Control and Disarmament Agency, and the Administrator of the National Aeronautics and Space Administration. Representatives of the Office of Management and Budget and the Office of Science and Technology Policy will be included as observers. Other agencies or departments will participate based on the subjects to be addressed.

BACKGROUND

In August 1981, the President directed a National Security Council review of space policy. The direction indicated that the President's Science Advisor, Dr. George Keyworth, in coordination with other affected agencies, should examine whether new directions in national space policy were warranted. An interagency working group was formed to conduct the study effort and Dr. Victor H. Reis, an Assistant Director of the Office of Science and Technology Policy was designated as Chairman. The group addressed the following fundamental issues: (1) launch vehicle needs; (2) adequacy of existing space policy to ensure continued satisfaction of United States civil and national security program needs; (3) Shuttle organizational responsibilities and capabilities; and, (4) potential legislation for space policy. The reports on the various issues formed the basis of the policy decisions outlined here. The following agencies and departments participated: State, Defense, Commerce, Director of Central Intelligence, Joint Chiefs of Staff, Arms Control and Disarmament Agency and the National Aeronautics and Space Administration, as well as, the National Security Council and the Office of Management and Budget.

Appendix 7
Hans Mark's Draft of Suggested Remarks for President Reagan, July 4, 1982, Edwards Air Force Base

Our country's birthday is always special but this one is unique. We have just witnessed a spectacular event—the fourth landing of "Columbia"—but what is more important, we have come to the end of the first period of this great enterprise and to the beginning of a new era in space.

On this flight, "Columbia" carried two payloads, one an experiment that is funded entirely by private industry and another one that is sponsored by the Air Force and is related in important ways to our national security. This illustrates the versatility of "Columbia" and it vindicates the original idea behind the Space Shuttle—which is that it is a space vehicle for all purposes. On the next flight, "Columbia" will conduct her first operational mission and will place two commercial communications satellites in orbit. From now on, "Columbia" and her sister ships will provide us with routine access to space for scientific exploration, commercial ventures and for tasks related to the national security.

I want to take this opportunity to congratulate Ken Mattingly and Hank Hartsfield on their achievement. It is in the great tradition of reaching for new horizons —something which Americans have always done superbly well. Lewis and Clark, Lindbergh and Byrd, Armstrong and Aldrin—these men are cut from the same cloth. We need not fear for the future of our nation as long as we have men such as these.

"Columbia's" fourth flight is just the beginning. The Space Shuttle gives us the means for establishing the permanent presence of mankind in space. This is the next step and there is no question that Americans will be in the lead. (I have, today, directed the National Aeronautics and Space Administration to initiate the first step toward the creation of a permanently manned American Space Station. This Space Station will fulfill the promise of the Space Shuttle and will clearly establish this country's pre-eminence in space well into the 21st Century.)

My fellow Americans, when I was inaugurated as your President 18 months ago I said that great nations must have great dreams. We are here witnessing the fulfillment of a great dream and this is evidence of what Americans can accomplish when we put our minds to it. I am proud to be here today to participate in the celebration of this great event.

Appendix 8
Text of Remarks by the President on the Landing of the Space Shuttle *Columbia*, July 4, 1982, Edwards Air Force Base

Dryden Flight Research Facility
Edwards Air Force Base, Calif.
July 4, 1982

In the early days of our republic, Americans watched Yankee clippers glide across the many oceans of the world, manned by proud and energetic individuals, breaking records for time and distance, showing our flag and opening up new vistas of commerce and communications.

Well, today you've helped recreate the anticipation and excitement felt in home ports as these gallant ships were spotted on the horizon, heading in after a long voyage.

Today we celebrate the 206th anniversary of our independence. Throughout history we've never shrunk before a challenge. The quest of new frontiers for the betterment of our homes and families is a crucial part of our national character—something which you so ably represent today. The space program, in general, and the Shuttle program, in particular, have gone a long way to help our country recapture its spirit of vitality and confidence. The pioneer spirit still flourishes in America.

In the future, as in the past, our freedom, independence and national well-being will be tied to new achievements, new discoveries and pushing back frontiers. The fourth landing of the Columbia is the historical equivalent to the driving of the golden spike which completed the first transcontinental railroad. It marks our entrance into a new era. The test flights are over, the groundwork has been laid, now we will move forward to capitalize on the tremendous potential offered by the ultimate frontier of space.

Beginning with the next flight, the "Columbia" and her sister ships will be fully operational and ready to provide economical and routine access to space for scientific exploration, commercial ventures and for tasks related to the national security. Simultaneously, we must look aggressively to the future by demonstrating the potential of the Shuttle and establishing a more permanent presence in space.

We've only peered over the edge of our accomplishments; yet already the space program has improved the lives of every American. The aerospace industry provides meaningful employment to over a million of our citizens—many working directly on the space program, others using the knowledge developed in space programs to keep us the world leader in aviation.

In fact, technological innovations traced directly to the space program boost our standard of living and provide employment for our people in such diverse fields as communication, computers, health care, energy efficiency, consumer products and environmental protection. It's been estimated, for example, that information from satellites saves hundreds of millions of dollars per year in agriculture, shipping and fishing.

The space shuttle will open up even more impressive possibilities, permitting us to use the near weightlessness and near-perfect vacuum of space to produce special alloys, metals, glasses, crystals and biological materials impossible to manufacture on Earth.

Similarly, in the area of national security our space systems have opened unique opportunities for peace by providing advanced methods of verifying strategic arms control agreements.

The Shuttle we just saw land carried two kinds of payloads: one funded entirely by private industry and the other, related to our national security, sponsored by the Air Force. This versatility of the Columbia and her sister ships will serve the American people well. Yet we must never forget that the benefits we receive are due to our country's commitment, made a decade ago, to remain the world leader in space technology.

To insure that the American people keep reaping the benefits of space and to provide general direction for our future efforts, I recently approved a National Space Policy Statement—which is being released today.

Our goals for space are ambitious, yet achievable. They include:

—Continued space activity for economic and scientific benefits.

—Expanding private sector investment and involvement in space-related activities.

—Promoting international uses of space.

—Cooperating with other nations to maintain the freedom of space for all activities that enhance the security and welfare of mankind.

—Strengthening our own security by exploring new methods of using space as a means of maintaining the peace.

There are those who thought the closing of the western frontier marked an end to America's greatest period of vitality.

Yet, we are crossing new frontiers every day, the high technology now being developed, much of it a by-product of the space effort, offers us and future generations of Americans opportunities never dreamed of a few years ago. Today we celebrate American Independence confident that the limits of our freedom and prosperity have again been expanded by meeting the challenge of the frontier.

We also honor two pathfinders. They reaffirm to all of us that as long as there are frontiers to be explored and conquered, Americans will lead the way. They and the other astronauts have shown the world that Americans still have the know-how and Americans still have the true grit that tamed a savage wilderness.

Charles Lindbergh once said that "short-term survival may depend on the knowledge of nuclear physicists and the performance of supersonic aircraft, but long-term survival depends alone on the character of man."

Appendix 9
Terms of Reference for the Space Station Study, April 1983

The White House
Washington

April 11, 1983

Memorandum for the Vice President
The Secretary of State
The Secretary of Defense
The Secretary of Commerce
The Director, Office of Management and Budget
The Director of Central Intelligence
The Chairman, Joint Chiefs of Staff
The Director, Arms Control and Disarmament Agency
The Director, Office of Science and Technology Policy
The Administrator, National Aeronautics and Space Administration

Subject: Space Station

The President has approved the attached National Security Study Directive on the Space Station.

For the President.

William P. Clark

The White House
Washington

National Security Study
Directive Number 5-83

April 11, 1983

Space Station

OBJECTIVE

A study will be conducted to establish the basis for an Administration decision on whether or not to proceed with the NASA development of a permanently based, manned Space Station. This NSSD establishes the Terms of Reference for this study.

GUIDELINES

The specific policy issues to be addressed are the following (responsible agencies are indicated in parenthesis):

—How will a manned Space Station contribute to the maintenance of U.S. space leadership and to the other goals contained in our National Space Policy? (NASA)

—How will a manned Space Station best fulfill national and international requirements versus other means of satisfying them? (NASA/State for national and international civil space requirements; DOD/DCI for national security needs.)

—What are the national security implications of a manned Space Station? (DOD/DCI)

—What are the foreign policy implications, including arms control implications, of a manned Space Station? (State/NASA/ACDA)

—What is the overall economic and social impact of a manned Space Station? (NASA/Commerce/State)

These five policy issues will be addressed for each of the four scenarios outlined below.

In order to assess the policy issues in a balanced fashion, NASA will provide a background paper outlining four example scenarios that represent possible approaches for the continuation of this nation's manned space program. These example scenarios are:

—Space Shuttle and Unmanned Satellites

—Space Shuttle and Unmanned Platforms

—Space Shuttle and an Evolutionary/Incrementally Developed Space Station

—Space Shuttle and a Fully Functional Space Station

A separate, unrelated, generic space requirements paper will be produced for use in addressing the national policy issues. The representative set of requirements for each space sector will be provided by DOD/DCI for national security and NASA/DOC for civil programs. A drafting group consisting of representatives of the DCI, DOD, DOC and NASA will coalesce the requirements into a single document. It will represent currently identifiable official agency statements of requirements for a Space Station. Long-term agency requirements and objectives should also be included.

IMPLEMENTATION

A Working Group under the Senior Interagency Group for Space has been established to conduct this study. The Working Group is chaired by NASA and includes representatives for DOD, DOC, DCI, DOS and ACDA. The Working Group will produce a summary paper that assesses the issues and identifies policy options. Results of the study will be presented to the SIG (Space) not later than September 1983 prior to presentation to the President. Papers produced by the Working Group will not be distributed outside the Executive Branch without the approval of the SIG (Space). The SIG (Space) may issue more detailed Terms of Reference to implement this study.

(Signed)
Ronald Reagan

Appendix 10
List of Participants and Agenda of
Cabinet Council Meeting, December 1, 1983

The White House
Washington

Cabinet Council on Commerce and Trade
December 1, 1983
2:00 P.M.
Cabinet Room

AGENDA
1. Combined Federal Campaign
2. U.S. Space Station Proposal (CM#434)

PARTICIPANTS
The President
The Vice President
Secretary Baldrige
Attorney General Smith
Secretary Clark
Secretary Hodel
Edwin Meese III
Director Stockman
Ambassador Brock
James Baker III
Jack Svahn
Deputy Secretary McNamar (Representing Secretary Regan)
Deputy Secretary Thayer (Representing Secretary Weinberger)
Under Secretary Ford (Representing Secretary Donovan)
Deputy Secretary Burnley (Representing Secretary Dole)
Deputy Director McMahon (Representing Director Casey)
William Niskanen (Representing Chairman Feldstein)
Richard Darman, Assistant to the President and Deputy to the Chief of Staff
Craig Fuller, Assistant to the President for Cabinet Affairs
Robert McFarlane, Assistant to the President for National Security Affairs
Larry Speakes, Assistant to the President and Principal Deputy Press Secretary
Lee Verstandig, Assistant to the President for Intergovernmental Affairs

Faith Whittlesey, Assistant to the President for Public Liaison
Wendell Gunn, Executive Secretary
Larry Herbolsheimer, Associate Director, Office of Cabinet Affairs

FOR PRESENTATION:
Verne Orr, Secretary of the Air Force
James Beggs, Administrator, NASA
Hans Mark, Deputy Administrator, NASA
John Hodge, Director, Space Station Task Force, NASA
Fred Khedouri, Associate Director, Natural Resources, Energy, and Science, OMB
Gilbert Rye, Director of Space Programs, NSC
Anthony Calio, Deputy Administrator, National Oceanic and Atmospheric Administration, DOC

ADDITIONAL ATTENDEES:
Michael Baroody, Deputy Assistant to the President and Director of Public Affairs
T. Kenneth Cribb, Assistant Counsellor to the President
Nancy Risque, Special Assistant to the President for Legislative Affairs
Daniel Murphy, Chief of Staff, Office of the Vice President

Index

Aaron, David, 76
Abrahamson, James A., 63, 131, 150, 216
Addabo, Joseph, 105
Agnew, Spiro T., 30
Air Force, 177, 178, 216; airlift capabilities, 94–95; Manned Orbiting Laboratory (MOL) 62, 63, 172; nuclear strategic deterrent forces, 94; operations in space for national security, 95–96; Reagan appointments, 118
Air Force Association, 96, 190
Air Force (magazine), 96
Air Force–NASA relations, 116, 130, 131, 175
Air Force–NASA symposium, 170–72
Air Force Program Acquisition Review (PAR), 130
Alaka, Daniel, 208
Aldridge, E. C. (Pete), Jr., 118, 177
Aldrin, Edwin (Buzz), 36, 37, 191
Algranti, Joe, 157, 160
Allen, H. J. (Harvey), 29
Allen, Lew, Jr., 20, 86, 87
Allen, Richard V., 109–10, 129–30, 131, 146
Allnutt, Robert F., 155
Alvarez, Luis, 15
American Association for the Advancement of Science, 201
American space program, 21, 47, 191, 208
American Vacuum Society, 58
Ames Research Center (NASA), 29, 30, 32, 36, 39, 43, 45–48, 50, 51, 53, 55, 56, 58, 61, 66, 92, 115, 126, 127, 135, 193, 215, 216
Anders, Bill, 30, 31
Anderson, Andrew, 86
Anderson, Kinsey A., 23
Anderson, Robert, 116
Andrews, Mike, 203
Andrus, Cecil, 76
Apollo space program, 3, 26, 30, 35, 36, 39, 49, 50, 53, 122, 128, 132, 182, 183, 191; *Apollo 8*, 30, 31; *Apollo 11*, 35, 36; *Apollo 13*, 42, 43; *Apollo 17*, 36, 44; Apollo-Soyuz, 50, 51, 68–69; Apollo 204 capsule, 223
Argus project, 17, 18, 19–20, 24, 60–61
Armstrong, Neil, 36, 37, 43, 123
Arnold, James, 139
Augustine, Norman, 162
Aviation Week, 129

Baker, Diane, 37
Baker, James A., III, 176, 178, 185, 190
Baker, William O., 56
Baker-Ramo advisory group, 56, 59, 60
Baldridge, Malcolm, 179, 185, 186

Barking Sands Air Force Station (Kauai Island, Hawaii), 25
Beggs, James M., 28–29, 30, 49, 115–16, 117, 121, 122, 126, 127, 128–29, 130, 131, 132, 133, 134, 135, 137, 140, 146, 147, 148, 149, 150, 152, 153, 156, 157, 158, 162, 163, 165, 174–83, 185, 187, 189, 190–93, 201–3, 207–8, 218, 220
Behrens, William W., III, 51
Berger, Wallace G., 166, 169, 203, 204, 205, 208
Berglund, Robert, 76
Berman, Alan, 114, 118
Berta, Michael, 130, 131, 146
Bing, George, 18
Bloom, Stewart D., 23
Bobko, Karol, 158
Bobko, Mrs. Karol, 158
Boland, Edward P., 105, 126, 206, 207, 208
Bolden, Charles E., 217
Bonestell, Chesley, 12
Bonney, Auriol Ross, 14
Borman, Frank, 30, 117
Boston University, 13
Boyd, John W., 216
Bradburn, David, 84
Branscomb, Lewis, 33
Branscomb committee, 33–34
Braun, Wernher von, 2, 12, 31–32, 39, 40–41, 42, 43, 44, 48, 51, 61–62
Brezhnev, Leonid, 83
Brickel, James, 106
British Interplanetary Society, 1
Brock, William, 186
Brookhaven National Laboratory, 16
Brown, George, 203
Brown, Harold, 15, 18, 60, 61, 66, 69, 71, 72, 76, 78, 87, 94, 95, 101, 102, 105, 109
Brzezinski, Zbigniew, 76
Bush, George, 126, 133, 184
Byrd, Richard Evelyn, 44

Cabinet Council on Commerce and Trade, 184, 190
California Institute of Technology, 16
Cannon, Lou, 179
Carey, William D., 201
Carter, James E., 60, 80, 83, 84, 94, 101, 102, 103, 109, 111, 129, 213
Carter administration, 92, 94, 117, 118, 122, 143, 144, 145–46, 164, 173, 181, 203
Casey, William, 176
Century Plaza Hotel (Los Angeles), 37
Cernan, Gene, 69
Challenger (shuttle), 151, 160, 205; accident, 216–24; External Tank, 216; O-rings, 218–22; Solid Rocket Motor (SRM), 216, 218–21
Chambers, Alan B., 56
Chapman, Dean, 47
Chayes, Antonia Handler, 65, 88
Chiles, Lawton, 202
Christofilos, Nicholas C., 16–17, 24
Chupp, Edward, 19
Clark, Joan, 154, 196
Clark, John, 42
Clark, William P., 120, 146–47, 163, 176–78, 180, 185
Clark University, 8
Clarke, Arthur C., 5
Claytor, W. Graham, 113
Cleator, P. E., 1–6, 8, 9, 26, 27; on nuclear energy, 4–6; problems of fuel, 3
Clements, William P., 120
Cochran, Jacqueline, 37
Cohen, Aaron, 48, 123
Colgate, Stirling, 15, 17–18, 19
Collier's, 12, 32
Collins, Mike, 37
Columbia (space shuttle), 36, 122–25, 126, 128, 129, 133, 188, 189; second flight (STS-2), 132, 133, 134; fourth flight (STS-4), 147–48, 150–51, 152, 173; technical problems, 90–91

Communications Satellite Corporation (COMSAT), 211
Congressional Affairs Office, 202, 203
Congressional Space Caucus, 208
Conrad, Pete, 50
Cooper, Robert S., 178, 200
Cortright, Edgar M., 42
Cosmos Club, 167, 181
Covault, Craig, 129
Crabbe, Buster, 177
Crippen, Robert, 123, 205
Criswell, David, 191
Critical Choices for Americans, Commission on, 55
Crow, Duward (Pete), 116
Culbertson, Philip E., 84, 105, 126–27, 131, 139, 146, 164, 174, 180, 192, 202, 209, 210
Currie, Malcolm, 65
Cutter, W. Bowman, 76, 102

Darman, Richard, 148–49
David, Edward, 45
Charles H. Davis Lecture, 118
Deaver, Michael K., 120, 134, 148
Defense Advanced Research Projects Agency, 178
Defense Intelligence Agency (DIA), 66, 176–77
Defense Science Board, 165
DeVries, Ralph D., 184
Discoverer (satellite), 62
Discovery, 58, 160
Domenici, Pete V., 201, 202
Donlan, Charles J., 32
DuBridge, Lee A., 30, 45
DuMond, J. W. M., 16
Duncan, Charles, 71
Dyna-Soar Program, 40, 62, 63

Echo (satellite), 62
Edwards Air Force Base, 69, 123, 124; STS-4, 147–48, 150–51, 152, 156

Eggers, Alfred J., 36
Eimer, Manfred, 26
Elam, Judy, 157
electrophoresis experiment, 173–74
Ellington Air Force Base, 151
Ellsworth, Lincoln, 44
Elms, James C., 48, 113, 114, 115–16, 117, 121, 131
Endicott, John, 119
Engle, Joe H., 134
Enterprise (space shuttle), 65, 66, 155–61, 180–81
European Space Agency (ESA), 154, 157, 159
Evans, Llewellyn J., Sr., 174
Evans, Llewellyn J. (Bud), Jr., 174–75
Evans, William, 86
Explorer I, 17, 18, 61
Explorer IV, 18, 19
"Extended Duration Orbiter," 198

Faget, Maxime, 39, 48, 93, 125, 175
Faget, Nancy, 175
Fanseen, James F., 133–34, 148, 150
Federal Aviation Administration (FAA), 115
Federal Emergency Management Agency (FEMA), 203
Fermi, Enrico, 27
Finarelli, Margaret (Peggy), 164–65, 172, 177, 184
Finch, Robert, 37
Finke, Wolfgang, 158
Finn, Terry, 164, 184, 203, 207
First Manned Orbital Flight (FMOF), 90, 93
Fletcher, James C., 41–42, 45, 47, 49, 92, 113, 116, 131, 138, 179–80, 191, 223
Fletcher space station committee, 138–39, 169, 181
Ford, Gerald R., 55, 56, 59, 60
Ford administration, 92, 120
Forrester, Jay, 52

Foster, John S., Jr., 15, 65
Frank, Nathaniel H., 20
Freiwald, Joyce, 208
Friedman, Herbert, 119
Frosch, Robert, 78, 80, 92, 93, 102, 103, 106–7, 112, 127
Fubini, Eugene G., 61, 165–66, 169
Fuller, Carolyn, 101, 209
Fuller, Craig, 148, 150, 153, 175, 177, 179, 184, 186, 203, 207
Fullerton, Gordon, 65
Fuqua, Don, 105, 107, 193

Gagarin, Yuri, 21, 22, 81
Garn, Jake, 204, 207
Garriott, Owen, 188, 189
Gemini program, 35, 39, 62, 63, 69
General Dynamics Corporation, 121, 218
Giacconi, Riccardo, 25
Gibbons, John H., 201–2
Gillam, Isaac T., 146
Gilruth, Robert R., 31–32, 36, 39, 48, 114, 116, 174
Gingrich, Newt, 119–20, 208
Glenn, John, 62, 83, 116, 159, 178, 205
Goddard, Robert H., 8–9
Goddard Space Flight Center, 42, 178, 205
Goett, Harry, 29
Goldstein, Howard, 47
Goldwater, Barry, 83, 116, 117, 126, 202
Goodall, Harry A., 68, 85
Goodman, Clark D., 11, 13
Goodwin, Glen, 47
Gordon, Frank, 19
Gorton, Slade, 202
Green, Bill, 207, 208
Griffin, Gerald D., 142, 175
Griffin, Richard L. (Larry), 157
Griffin, Sandy, 175
Grumman Aerospace Corporation, 174
Guernsey, Janet B., 11
Gutman, Richard, 73

Hackerman, Norman, 60
Hahn, Otto, 5
Haize, Fred, 43, 65
Harper, C. W. (Bill), 29
Harper, Edwin, 130
Hart, Gary, 100–101
Hart, Terry, 205
Hartinger, James V., 88, 171
Hartsfield, Henry, 150, 151
Heflin, Howell, 202
Heiss, Klause P., 49
Helman, Gerald, 153
Helmholz, August Carl, 10
Helms, Jesse, 164
Henson, Carolyn, 54
Henson, H. Keith, 54
Herman, Daniel H., 139–40, 164
Hermann, Robert J., 88–89
Herres, Robert, 63
Hess, Wilmot N., 36
Hewitt, Jack, 65
Hildebrand, Joel H., 10
Hill, James E., 70, 71, 85, 88
Hill, Jimmie D., 68, 171
Hodge, John D., 140, 164, 203, 210
Hoffman, Hans, 154
Hornby, Harry, 47
Houbolt, John, 35–36
Hoyt, Diana, 208
Huberman, Benjamin, 79, 102
Huddleston, Walter, 101, 209
Hughes, Philip, 177
Hutchinson, Neil, 210

Interagency Group (Space), 172–73
Intercontinental Ballistic Missiles, 119
International Geophysical Year, 16

Jackson, Henry M., 25
Jackson, Roy P., 46, 90
James, Daniel E. (Chappie), Jr., 70
James, Pendleton, 121, 122
Jay, John, 55
Jefferson, Thomas, 213–14

Jenkin, Patrick, 160
Jenkin, Mrs. Patrick, 160
Jet Propulsion Laboratory, 26
Johnson, Lyndon Baines, 34
Johnson, Richard, 193
Johnson, U. Alexis, 30
Johnson and Johnson, 174
Johnson Space Center, 39, 48, 56, 122, 127, 128, 129, 132, 133, 136, 138, 140, 141, 151, 157, 169, 174, 175, 188, 189, 192, 203, 210, 212, 217
Joint Institute for Laboratory Astrophysics, 33
Jones, David C., 69, 85, 86
Jones, Tom, 116
Jopson, Robert M., 19, 23

Kaplan, Joseph, 12
Karlsruhe University, 14
Keeney, Spurgeon, 82
Kempis, Thomas à, 197
Kennedy, John F., 21–22, 35–36, 45, 48, 120, 132, 182–83
Kennedy Space Center, 64, 128, 129, 132, 150, 156, 188, 212, 218
Kerwin, Mary D., 203
Keyworth, George A., II, 130, 143, 144, 145, 146, 149, 167, 182, 183, 185, 191, 193, 198, 201
Khedouri, Frederick N., 170, 181, 182, 190
Kiehn, Robert M., 11
Kirtland Air Force Base, 20
Kissinger, Henry, 50
Koehler, John, 104
Kohl, Helmut, 189
Kraft, Christopher C., 48, 57, 58, 122–23, 133, 134, 141
Kranz, Gene, 132
Krebs, Tom, 177
Kubasov, Valery N., 51

Lance, Bert, 71
Langley Research Center, 22, 32, 35, 42, 48, 115, 184
Larson, Howard, 47, 90
Lawrence, Ernest O., 10, 15
Lawson, Robert D., 14
Leonov, Aleksey A., 51
Lewis, Harold, 11
Lewis, Jerry, 207
Lewis Research Center, 32, 115, 208
Ley, Willy, 6–8, 12, 27, 28
Lichtenberg, Byron, 188, 189
Lilly, William E., 92–93, 102
The Limits of Growth (Meadows, Meadows, Randers, Behrens), 51–52
Livermore National Laboratory, 15–20, 22, 24, 25, 26, 27, 29, 61, 65, 120, 182; Livermore P-Division, 20, 22, 23, 24, 26, 28, 61
Lockheed Corporation, 193
Los Alamos National Laboratory, 27, 143
Louis, John, 160
Louis, Mrs. John, 160
Lovelace, Alan, 78, 90, 93, 102, 106, 112, 125
Lovell, Jim, 30, 43
Low, George M., 35, 38–39, 42, 45, 48, 50, 78, 117, 121, 142
Lujan, Manuel, Jr., 208
Lundin, Bruce T., 35

McCartney, Forrest S., 171
McClelland, Clyde L., 11
McConnell, Nadia, 203
McDonnell Douglas Corporation, 47, 58, 174
McFarlane, Robert C. (Bud), 153, 186
McIlwain, Carl, 18
McIntyre, James, 71, 72, 102, 103
McKay, Gunn, 101
McLucas, John L., 120
McMahon, John, 177
McMillan, Edwin, 15
McMurtry, Tom, 157
McNamara, Robert, 40, 63
"McNeil-Lehrer Report," 110

Madison, James, 55
Malone, James, 163, 181
Malow, Richard N., 206, 207, 208
Mann, Lloyd, 23
manned space flight, 40, 42, 43, 127
manned vs. unmanned space flight, 34, 50, 63, 138–39, 144–45, 165–67, 172, 188–89, 198, 204, 205, 206
Mark, Herman F., 9, 10, 14
Mark, Marion T. (née Thorpe), 11, 13, 14, 114, 157
Mars, 33, 34, 42
Marshall Space Flight Center, 31, 39, 62, 127, 136, 137, 138, 220
Martin, John J., 65, 88
Massachusetts Institute of Technology (MIT), 11–13, 14, 20
Mathematica (consulting firm), 49
Mathias, Charles, 105
Mattingly, T. K., 150, 151
May, Michael M., 15
Mayo, Robert P., 31
Meadows, Dennis H., 51
Meadows, Donella H., 51
Meese, Edwin, III, 153, 176, 185, 190
Mendell, Wendell, 170
Merbold, Ulf, 188, 189
Mercury program, 21, 39, 63, 141, 183
Merkle, Theodore, 15
Messerschmitt-Boelkov-Blohm (MBB), 159
Metzger, Alfred, 26
MIRAK (Minimum Rakete), 2, 6, 8
Mission Control Center (MCC), 36, 123, 125, 133–34, 135, 150, 151, 188
MIT. See Massachusetts Institute of Technology
Mondale, Walter, 103
Morgenstern, Oskar, 49
Morgenthaler, George, 40
Morris, Jay, 117
Mueller, George E., 31–32, 34, 40, 42, 47, 114
Mueller OMSF management council. See Office of Manned Space Flight management council

Murphy, Daniel J., 79, 173
Murphy, John F., 202
Myers, Dale, 42, 47

National Aeronautics and Space Act of 1958, 64, 75, 182, 199
National Aeronautics and Space Administration (NASA), 21–22, 23, 26, 28, 31, 39, 41–42, 45, 47, 48, 49, 53, 57, 58, 61, 62, 63, 66, 92, 93, 116, 119, 125, 126, 127, 128, 129, 133, 138, 145, 162, 163, 164, 166, 167, 169, 170, 172, 175, 177, 178, 182, 183, 184, 185, 191, 192, 193, 195, 196, 197, 198, 199, 200, 202, 203, 208, 210, 211, 212, 215, 216, 218, 219, 221, 222, 223; Air Force agreement, 65; and Air Force relations, 116, 130, 131, 175; and Air Force space collaborations, 64–65; Appropriations Subcommittee, 204; budget, 190, 201; Civil Service, 128; development of manned station, 139; Electronics Research Center, 115–16; Office of Congressional Affairs, 202; problems with Department of Defense, 130; Reagan administration's "transition team," 117, 121; role, in shuttle Orbiter, 64; securing a post under Reagan, 120–21; visit by Reagan's senior staff, 153
NASA Aeronautics and Astronautics Coordinating Board, 65
NASA–Air Force symposium, 170–72
NASA's Ames Research Center. See Ames Research Center
NASA's International Affairs Office, 164
NASA's Lewis Research Center. See Lewis Research Center
NASA's Manned Spacecraft Center (now Johnson Space Center), 32, 36
NASA's Planetary Program Office, 139
National Air and Space Museum, 179, 182
National Science Board, 60

National Science Foundation, 59, 60
National Security Council, 130, 131, 146, 153
Naval War College, 118
Nedzi, Lucian, 101
Nelson, George D. (Pinky), 205
Neutron Physics Group, 11, 13
Newell, Homer E., 31, 38
Nixon, Richard M., 30, 36, 37–38, 45, 47, 55, 56, 59, 92, 115, 125, 132, 180, 183
Nixon administration, 47, 116
North American Air Defense Command (NORAD/ADCOM), 70
"Notes on Long Range Planning" (Mark, Silveira), 127–29
Nuclear test Orange, 24
Nuclear test Starfish, 24
Nuclear test Teak, 24

Oak Ridge National Laboratory, 27
Oberth, Hermann, 1–2, 6, 7
O'Brien, Morrough P. (Mike), 20
Office of Management and Budget (OMB), 67, 71, 92, 145, 149, 152, 170, 171, 181, 185, 186, 190, 192
Office of Manned Space Flight (OMSF), 40
Office of Manned Space Flight (OMSF) management council (Mueller), 31–32, 34, 42
Office of Science and Technology Policy, 59, 143, 149, 182, 184
Office of Technology Assessment (OTA), 201–2, 206
O'Neill, Gerard K., 51–54
Orbiter, 48, 64, 66, 72, 152, 171–72, 197–98, 200
Orbiter Project Office, 132
Orr, Verne, 88, 118, 129, 130, 178, 200

Paine, Thomas O., 30, 38
Paris Air Show (Le Bouget airfield, Paris), 155–56, 160
Parker, Robert, 188, 189

Payload Operations Control Center (POCC), 188
Perry, William J., 71, 72, 94, 101, 102, 113
Petersen, Richard, 47
Peterson, Vic, 47
Physics Today, 53
Pioneer series, 139
post-Apollo program, 30–31, 32, 35, 37, 39, 41, 63
Pournelle, Jerry, 191
Powell, Luther, 164
President's Advisory Group on Science and Technology, 56, 61
President's Scientific Advisory Committee (PSAC), 59
Press, Frank, 77, 78, 80, 102, 103, 107, 146
Project Argus. See Argus project
Proxmire, William, 95, 105

Die Rakete zu den Planetenraumen (Oberth), 1
Ramo, Simon, 56
Randers, Jorgen, 51
Randolph Air Force Base, 216
Ranger (unmanned spacecraft), 26
Reagan, Nancy, 150
Reagan, Ronald, 8, 79, 82, 108–10, 111, 113, 118, 120, 133, 135, 148, 149, 150, 151, 152, 163, 173, 175, 176, 177, 178, 179, 182, 183, 184, 185, 186, 190, 191, 192, 193, 194, 196, 197, 199, 207, 214, 215, 217, 218; at launch of Columbia, 134; personal interest in space program, 131, 151, 152, 153, 165
Reagan administration, 50, 116, 119, 143, 144, 164, 198, 201; position on shuttle, 129, 130; space policy, 145, 146
Redstone Arsenal (Huntsville, Alabama), 62
Reed, Thomas C., 120
Reis, Victor H., 143, 146

Reisenhuber, Heinz, 158, 180–81
Relay (satellite), 62
Rensselaer Polytechnic Institute, 117
Rickover, Hyman G., 11
Roberts, Leonard, 47
Rockefeller, Nelson, 54–55, 56, 59, 60
Rockets and Space Travel (Ley), 6–8
Rockets Through Space (Cleator), 1–6, 9
Rogers, Craven C., 118
Rogers, William P., 218
Rogers Commission Report, 218, 220, 221
Roosevelt, Franklin D., 194
Rosenberg, Robert A., 77, 79, 130
Rostow, Eugene, 164
Ruebhausen, Oscar, 56
Rutherford, Lord, 5
Ryan, Cornelius, 12
Rye, Gilbert D., 146, 147, 149, 163, 165, 172, 184, 185, 190, 207

SALT II arms control treaty, 81, 83
Salyut (space station), 185
Sandler, Harold, 56–57
Saturn V, 3, 62
Saunders, Harold, 82
Sawyer, Thomas W., 85
Schirra, Wally, 69
Schmitt, Harrison (Jack), 36, 100, 105, 117, 122, 179
Schneider, William, Jr., 130, 177, 181
Schrader, Carleton D., 26
Schriever, Bernard M., 62, 71, 116
Science Advisory Committee (Nixon), 166
Science (magazine), 146
Scobee, Dick, 157
Score (satellite), 62
Seaborg, Glenn T., 15, 30
Seamans, Robert C., 30
Seitz, Frederick, 57, 216
Senior Interagency Group (Space) (SIG [Space]), 146, 147, 153, 163, 164, 172–73, 177, 178
Seward, Frederick D., 24–25

Sharp, Joseph C., 56
Shepard, Alan, 178
Shuttle Orbiter Project Office (JSC), 123, 127
Sidey, Hugh, 22, 183
Silveira, Milton, 48, 123, 127, 141, 175
Silver, Samuel, 28
Skylab, 50
Slay, Alton D., 69, 87
Smith, Donald D., 117
Smith, Larry K., 100
Smith, William French, 185, 186
Solar Maximum Mission (SMM) satellite ("Solar Max"), 141, 205
Spacelab program, 140, 154, 159, 188, 189
Space Shuttle, 219, 222
space shuttle program, 32, 33, 34, 39, 40, 41, 45, 47, 48, 51, 57, 58, 62, 64, 67, 96, 127, 128, 130, 132, 140, 144, 153, 180, 183, 191, 192, 197, 198, 206, 211, 212, 213, 214; costs, 49–50; divided between NASA and USAF, 64–65; economic and political justifications, 49; *Enterprise* as U.S. exhibit centerpiece in Paris Air Show, 155; first Approach and Landing Test (ALT), 65, 66, 155; funding problems, 92–93; operations moved from JSC to KSC, 128, 129; Reagan's personal interest, 131; reusable thermal protection system, 42, 47
space program: Reagan's personal interest, 151, 152, 153; sequence of shuttle and space station, 41
space station, 1, 2–3, 32–34, 40, 51–54, 127, 137, 140–41, 144, 146–47, 152, 158–59, 169, 170, 173, 176, 177, 179, 183, 189, 190, 191, 193, 198, 199, 200, 202, 203, 206, 208, 209, 210, 212; appearance, 136; collaboration between United States and Western Europe, 180–81; convincing Congress to approve, 205; and defense against

strategic weapons, 200; Fletcher committee options, 138; gaining support for, 161, 162; Germany's support, 180–81; and the "interagency process," 163–64, 165, 172, 177, 185; as international effort, 192, 203–4, 207; opponents of, 162, 165–67, 171–72, 176, 177, 184, 201; permanently orbiting manned, 128, 179, 182, 199, 210, 215; persuading the Reagan administration of its feasibility, 132, 136, 143, 147, 148, 184; Reagan's speech at STS-4, 149; and the Russians, 185; at SIG (Space) meeting, 157; in State of the Union address, 194–96; supporters of, 178, 186
Space Station Task Force, 164
Space Task Group (STG), 30–31, 33, 34, 38–39, 41
Space Technology (magazine), 129
Sputnik I, 18, 21, 182
Stafford, Thomas, 51, 68–69, 97, 116, 118
Stetson, John C., 65, 86, 87
Stevenson, Adlai E., III, 101
Stever, H. Guyford, 59, 60
Stilwell, Richard, 172
Stine, G. Harry, 191
Stockman, David A., 170, 185, 186, 190
Stokes, Louis, 208
Strassmann, Fritz, 5
Strategic Defense Initiative Office, 216
Strub, Herman, 158
Studdert, Steve, 134
Submarine Launched Ballistic Missiles (SLBM), 119
"Super PAR," 130–31
Surveyor (unmanned spacecraft), 26
Swenson, Byron, 47
Swift, Charles D., 23
Swigert, Jack, 43
Syncom (satellite), 62

Talley, Wilson K., 116
Tate, Thomas, 193–94

"Technology and the Strategic Balance" (Mark), 118–19
Teller, Edward, 14, 54–55, 116, 167, 170
Terrell, Norman, 164, 184
Thayer, Paul, 163, 186
Thiokol, 220, 221, 222
Tilling, Reinhold, 2
Time (magazine), 22, 183
Titan, 40, 62, 172
Tower, John, 116
Tracking and Data Relay Satellite (TDRS), 188
Transit (satellite), 62
Transportation Systems Center— Department of Transportation, 116
Trible, Paul, 202
Trombka, Jacob, 26
Truly, Richard H., 134, 223
Turner, Stansfield, 71, 72, 77

UNISPACE 82 (UN Congress on Space), 153–54, 156
U.S. Air Force, 61–64; role in shuttle Orbiter, 64, 65
U.S. Arms Control and Disarmament Agency (ACDA), 164
U.S. Army's Ballistic Missile Agency (AMBA), 12, 22, 62
U.S. Army's Jupiter Intermediate Range Ballistic Missile (IRBM), 61
U.S. Atomic Energy Commission, Nuclear Test Site, 22
U.S. Department of Defense, 130, 145, 178, 179, 186, 216
U.S. Department of Transportation, 115–16, 140, 196
U.S. House and Senate Appropriations Committee, 141
U.S. House Science and Technology Committee, 193, 198, 200, 201, 202, 203
U.S. House Subcommittee on Space, Science and Applications, 203
U.S. Naval Research Laboratory, 16,

118, 119, 121
U.S. Senate Budget Committee, 201
U.S. Senate Subcommittee on Housing
    and Urban Development—
    Independent Agencies, 204
U.S. Senate Subcommittee on Science,
    Technology, and Space of the Com-
    mittee on Commerce, Space, and
    Transportation, 122
University of California (Berkeley), 10,
    14, 15, 20, 28, 34
University of California's Space Sciences
    Laboratory, 28
University of Texas System, 215, 216
unmanned space operations, 40, 45

V-2 rocket, 6, 12
Valier, Max, 2
Van Allen, James A., 17–18, 19
Van Allen Radiation belts, 17, 19, 23
Van de Graaff, Robert J., 11
Vandenberg Air Force Base, 25, 64, 96,
    129, 130–31, 212
"Vanguard," 16
Van Hoften, James D., 205
Verein für Raumschiffahrt (vfr), 2, 8
Viking project, 42, 139
Volkmer, Harold, 200
Volpe, John, 116
Von Pirquet, Count Guido, 2, 7, 44
Voyager, 139

Walker, Charles, 58
Warnke, Paul, 82
Washington Post, 117, 178–79
Washington Space Station Program
    Office, 192

Watergate, 55–56
Watkins, James, 176, 177
Watson, Jo, 118
Watt, James, 180
Webb, James E., 22, 191
Weeks, L. Michael, 112, 219, 221
Weinberger, Caspar, 172, 173, 175–76
Weiss, Stanley, 112
Weisskopf, Victor F., 12, 14
Weitz, Paul, 50
Westinghouse Corporation, 115
Wexford estate (Middleburg, Virginia),
    120
Wheelon, Albert D., 108–9, 110, 116
Whitehead, Clay T. (Tom), 110
White House Legislative Affairs, 204
White House Program Resources Board,
    190
White House Science Council, 167
Wiesner, Jerome, 21, 183
Wiesner report, 21
Williams, General (DIA), 176
Williamson, David, 92
The Window of Opportunity (Gingrich),
    120
Winkler, Johannes, 2
Winters, Jonathan, 37
Wood, Lowell, 191
Wright, Joseph, 153, 177

X-24B, 40
X-rays, 23–25

Yardley, John F., 47–48, 90, 102, 112
York, Herbert, 15, 19, 61
Young, A. Thomas, 114
Young, John, 69, 123